Embargoed Science

Embargoed Science

VINCENT KIERNAN

UNIVERSITY OF ILLINOIS PRESS

Urbana and Chicago

Library of Congress Cataloging-in-Publication Data
Kiernan, Vincent.
Embargoed science / Vincent Kiernan.
p. cm.
Includes bibliographical references and index.
ISBN-13: 978-0-252-03097-0 (cloth : alk. paper)
ISBN-10: 0-252-03097-4 (cloth : alk. paper)
1. Science news. 2. Reporters and reporting.
3. Journalism—Social aspects.
I. Title.
Q225.K48   2006
500—dc22   2006000323

To Terri, Emily, and Matthew,
who never embargo their love for me

Contents

Acknowledgments

When I tiptoed into this research project, the late Marjorie Ferguson encouraged me to scrutinize journal embargoes. Although I encountered uninterest or even hostility in other quarters, her support helped me to stay on track, even after she fell ill. Into the void left by Marjorie's death stepped Kathy McAdams, and I could not have asked for a more helpful scholar, armed with a discerning eye and ready encouragement. I also am grateful for the assistance and guidance I received from Maureen Beasley, Marcel LaFollette, Bruce Lewenstein, and Carol Rogers. Kerry Callahan at the University of Illinois Press was a source of excitement and encouragement, and Annette Wenda read the manuscript with a careful eye. Deborah Blum, Boyce Rensberger, and Mark Jerome Walters provided valuable critiques. My colleagues Jody Brannon, Michael Dorsher, John Cordes, and Linda Hagan all helped me shape my research in ways large and small.

In addition, many people assisted me in the collection of the data for this project. Officials from four journal publishers provided crucial information for the content analysis regarding the participation of various newspapers in their embargoes: Nan Broadbent and Ginger Pinholster of the American Association for the Advancement of Science; Jerome P. Kassirer, formerly of the *New England Journal of Medicine;* Laura Garwin of *Nature;* and Jeff Molter of the American Medical Association. Some of these individuals—perhaps all of them—may disagree strongly with the conclusions that I draw in this book, but I hope that they do not regret their cooperation with me.

Embargoed Science

1

An Overview of News about Science and Medicine

Black holes in distant space. The latest AIDS treatments. Evidence that pesticide residues in food cause cancer—or don't. Quarks, leptons, and cold fusion.

On any given day, the news media may cover these or other news stories on a variety of topics about scientific and medical research. This chapter will describe the processes by which this science and medical news is socially constructed and used. First, I will describe the media organizations and media professionals who prepare and deliver news about science and medicine. Then we will consider the communication systems in science that form the basis for much journalism about science and medicine.

Science and Medical Journalism

At many media organizations, journalists are given responsibilities for covering specific topics. These arrangements of coverage responsibility are known as "beats." Usually, beats are defined bureaucratically. In most cases, this bureaucratic orientation is explicit: the White House beat, in which reporters monitor the activities of the holder of one specific elected office, is an excellent example.

The beat reporter is expected to monitor developments on the beat and identify stories as they arise. The beat arrangement allows the reporter to accumulate expertise in a field and build a larger stable of sources who can provide information for stories. In this sense, the beat system is designed to efficiently collect bureaucratically organized news.[1] Some reporters have no specific beats, and may be designated as "general assignment" or "national" reporters, which means that they are assigned stories by editors and that the topics of those stories may vary widely from one day to the next. Much of the science and medical reporting in the news media is produced by journalists who are assigned to science and medical beats. As of 2005, the National Association of Science Writers (NASW) included about 1,400 journalists at newspapers, magazines, and radio stations and in television. The Association of Health Care Journalists had about 725 journalists as members as of mid-2004.[2] However, membership in such groups is at best a rough approximation to the population of U.S. science and medical journalists, as some journalists in the field—even at major media organizations—are not members.

Among the various news media, newspapers clearly are the dominant force in disseminating news about science and medicine. More so than other media outlets, newspapers devote staff and news space to science and medicine. Even so, the coverage is less than many scientists might like: for example, one recent analysis of the content of fifteen U.S. newspapers of varying sizes, for selected dates in 2003, found that about 5 percent of the front-page stories dealt with science.[3] Some of the largest newspapers, such as the *New York Times* and the *Los Angeles Times,* have substantial staffs of science and medical reporters, each of whom specializes in certain disciplines. Smaller newspapers may have only one reporter who covers both science and medicine. Many smaller newspapers have no one assigned to the science and medicine beat.

About a decade ago, one public-relations directory suggested that, in the United States, few newspapers with circulations of less than fifty thousand had any reporter designated to cover science or health.[4] Similarly, this directory suggested, few radio stations had science or medical reporters, even part-time, and only local television stations in larger markets were likely to cover medicine or science substantively. This description remains apt. Radio is largely devoid of reporting on science and medicine, with the notable exception of National Public Radio; as one observer has described the situation, "Science appears on most commercial radio stations only as bits of news taken from the syndicated news services."[5]

Commercial broadcast journalism—both at the network level and at local stations—largely eschews science reporting, although reporting about health and medicine is more popular. "There is very little science news on TV . . . and most of what you get is in the areas of health and medicine," says Ira Flatow, a veteran science broadcast journalist. To redress that problem, he helped start a company, now called ScienCentral, that supplies television networks and local stations with short video news items about science. Although the broadcast networks at one time had separate medical and science reporters, the networks have combined the duties into that of a single merged health-science reporter.[6]

The prevalence of health and medical coverage in local television news programming has been bolstered by news consultants who have gained increasing influence over local news content as station owners seek to maximize their profits. In one 2001 survey of news directors of local television stations, 23 percent reported that consultants had encouraged them to broadcast coverage of health and medicine, although this clearly does not necessarily mean coverage of medical research. The same study examined news programming in 124 cities in 2001 and found that about 11 percent dealt with science and technology. The importance of medical reporting was also evident from allocation of reporting resources; in a 2002 survey of news directors at local stations, 42 percent related that their reporting staffs included a full-time reporter covering health or medicine.[7]

But the reporting nevertheless is superficial. As one former CNN medical correspondent notes: "Much local TV health and medical coverage looks like the media equivalent of a 99-cent drive-thru menu: quick, cheap, but ultimately unnourishing." An editor at National Public Radio described health reporting on local television as dominated by journalists who are incapable of independently assessing published research: "If there's a story in today's *Journal of the American Medical Association,* you take a look at it, and you call the local hospital, even though they didn't do the research (you never talk to the researcher). They explain it after they look at it, you do two sound bytes and a little cover in between. You look for a patient who has this problem. . . . You always have to start with an anecdote."[8]

At the network level, science news also receives little attention. A content analysis of both evening and morning network news broadcasts in 2003 found that science was the topic of 2 percent of each type of broadcast. (At first blush, this figure would appear to be not markedly lower than the 5 percent of newspaper stories devoted to science. However, one must bear in mind that the typical television news broadcast contains fewer words than a newspaper's front page. Consequently, the total amount of television science coverage is far less than the total amount of newspaper science coverage.)

The study also found that the evening and morning broadcasts took different approaches to covering science news. The morning shows are "where science is covered as innovations in personal health or consumer electronics; and where environmental stories such as global warming are covered as the latest weather disaster." A 1988 content analysis of four days of evening programming by ABC, CNN, NBC, and PBS showed that science—which the study defined to exclude medical research—received less broadcast time at each network (or was tied for last place) than eleven other news categories. The one exception was CBS, for which science was the sixth-ranked topic during those four days. A similar scarcity of science reporting was documented in another content analysis of the evening news broadcasts by the three major networks, over a ten-week period in 1997, which found that 4.4 percent of the stories broadcast by the three networks were on science, technology, or computers.[9] The content analysis described in the Appendix, which included three commercial networks' evening news broadcasts from June 1997 through May 1998, found a total of sixty-two reports on research published in the *Journal of the American Medical Association, Nature,* the *New England Journal of Medicine,* or *Science;* the two medical journals received the bulk of the attention, with each being covered by twenty-seven network newscasts.

One observer has suggested that efforts to promote the public understanding of science should focus on cable television networks, because of these channels' popularity and viewership growth. But at present, cable news offers little coverage of science. CNN, Fox, and MSNBC each devoted 1 percent of their programming to science in 2003—equivalent to two and a half minutes in one sixteen-hour period. An additional thirty seconds was devoted to medical research. "As for science news on cable TV, it is quite rare. Yes, there are cable channels that profess to carry science, but very little science news really shows up. Most of what shows up are animal stories, gadgets, emergency room operations and Internet how-to's," Flatow says.[10]

In the arena of print journalism, newsmagazines' medical coverage is focused on "news you can use" rather than medical research.[11] Science-oriented magazines fall into two categories: small magazines that focus on specific subjects and a small number of large generalist magazines. Several popular-science magazines that started in the 1980s faltered due to an inability to attract advertising.[12] Indeed, newspapers appear to be the primary medium for disseminating popular news about research in science and medicine. Marianne Pellechia examined science coverage in three major American daily newspapers in three time periods—the late 1960s, late 1970s, and late 1980s—and found that science news accounted for an increasing proportion of the content of the newspapers. In all three time periods, medicine and health accounted for more than 70 percent of the science stories in the three newspapers, with the rest split in varying proportions between technology and natural and physical sciences.[13]

Medical news appears to dominate science coverage in newspapers published outside the United States as well. A content analysis of Canadian daily newspapers in 1986 and 1987 found that 45 percent of science stories were on medicine or health; the next most common category, environmental stories, accounted for only 15 percent.[14] Some newspapers devote special weekly sections to science news. Such sections became more prevalent in the 1980s, fueled in part by growth in advertising related to personal computers. But science sections fell into decline in the 1990s, as advertising became more scarce.[15]

The World Wide Web is another medium for disseminating news about science and medicine. However, in many cases, the Web sites operated by media organizations offer little original content, instead presenting Associated Press (AP) stories and content from the newspaper or other media organization that sponsors the site.[16] Although science-communication researchers have left the science-news content of news-media Web sites largely unscrutinized, one recent content analysis compared the coverage of cloning from 1996 through 1998 by ABC.com, CNN.com, and MSNBC.com with coverage of that issue by the *New York Times,* the *Washington Post,* and *USA Today.* Online articles were shorter. The Web sites largely relied on content provided by wire services, whereas wire-service copy represented only 6 percent of the newspaper articles. But both took a positive, forward-looking tone to their coverage.[17]

Science news is also available through stand-alone Web sites. One of the best known is Why Files (http://whyfiles.org), a freely available Web site operated by the graduate school of the University of Wisconsin at Madison that attempts to explore "the science behind the headlines," with amusing articles that are easily accessible to the public.[18]

Coverage of issues in science and medicine is, of course, not limited to reporters on those two beats. Environmental reporters, for example, often cover issues related to chemistry, toxicology, and biology. Reporters covering criminal prosecutions may have the challenge of covering DNA evidence. Business reporters cover biotechnology and high-technology computing. Even sports reporters may have to cover science-based issues such as those relating to drug use by athletes. But science and medical journalists usually have a virtual monopoly on covering scholarly journals; a media organization may miss covering locally produced research published in a major scientific journal if its science reporter is unavailable.[19]

Also, media organizations are not limited to the stories that their own reporters can generate. They also use content that has been produced by journalists at other media organizations. The most prominent such source is the

Associated Press, a cooperative of news organizations that disseminates both stories that have been written by its member organizations and stories that have been written by the AP's own staff reporters. The AP distributes newspaper stories, photographs, video, and audio, and it supplies news content for Web sites; it has been described as the "backbone" of American journalism.[20] The AP has several reporters who cover science or medicine full-time.[21] The wire service is a dependable supplier of regular science news to media organizations: from June 1997 to May 1998, for example, the Associated Press distributed 682 breaking-news stories covering the *Journal of the American Medical Association, Nature,* the *New England Journal of Medicine,* and *Science.*[22] Although major newspapers might be expected to eschew the use of wire services because they have staff reporters who specialize in covering science and medicine, these media outlets nevertheless rely heavily on the wires: one content analysis of 1988 coverage of the *Journal of the American Medical Association* and the *New England Journal of Medicine* by ten major U.S. daily newspapers found that 38 percent of the articles on these journals bore a wire-service byline. Smaller media organizations that do not have science or medical reporters are likely to rely even more heavily on wire services for their science and medical news. One analysis of more than 80,000 health-related stories in four midsize Missouri newspapers found that 54 percent came from wire services and news syndicates.[23]

Although media organizations rely heavily on the Associated Press, wire copy may differ from staff-written copy. Because wire articles are written for a national audience, by definition they ignore local aspects of science and medical stories to which local media would be expected to devote attention. But there may be other differences as well. One content analysis—which compared coverage of the search for life on Mars by the *Viking* spacecraft in 1976 by the *New York Times,* the *Washington Post,* the Associated Press, and United Press International (UPI)—found that newspaper articles were more complete, although the difference may have been attributable in part to the fact that newspaper articles also were longer than wire stories.[24]

Media organizations can also draw on resources beyond the AP. The overwhelming majority of daily newspapers are no longer family owned but rather are the property of media corporations. Many of these chains encourage their newspapers to share stories with one another, and some of these chains also sell syndicated access to the articles to media organizations that they do not own. Examples of such services that often carry science news include the New York Times News Service and the Knight Ridder/Tribune Information Services.

Scientists often complain that science journalists have little formal scientific training, but current data on this question are sparse. One recent review of science-news coverage states: "Journalists tend not to have even a liberal-arts background in the sciences. Few understand the scientific method, the dictates of peer review, the reasons for the caveats and linguistic precision scientists employ when speaking of their work."[25]

Science and medical journalists have tended to be better educated than other journalists, both in general and in terms of science courses. One 1978 survey of science journalists found that 29 percent had graduate degrees; by contrast, a 1983 survey of journalists as a whole indicated that 11 percent had graduate degrees. A 1991 survey of 108 medical journalists found that 50 percent had a bachelor's degree, 36 percent had a master's degree, 9 percent had doctorates, and 2 percent held medical degrees. A 1993 survey of 83 science reporters at American daily newspapers of circulation exceeding 100,000 found that all had at least a bachelor's degree, with 73 percent of them in journalism. Twenty-two percent held a master's degree, and 1 respondent had earned a doctorate.[26]

However, the educational gap between generalists and science and medical journalists may be narrowing. A 1999 survey of 165 health-care reporters in the midwestern United States found that 84 percent had earned at least a four-year college degree: 67 percent held a bachelor's degree, 16 percent had a master's degree, and 1 percent had a doctorate. Those results are consistent with a 2002 survey of print and broadcast journalists, in which 89 percent reported having at least a four-year degree.[27]

Some science or medical reporters have extensive training in science or medicine. For example, Joseph Palca, a science reporter for National Public Radio, holds a doctorate in physiological psychology. Sue Goetinck Ambrose, of the *Dallas Morning News,* has a doctorate in molecular genetics. Several physicians are medical reporters, such as Lawrence K. Altman of the *New York Times,* David Brown of the *Washington Post,* Timothy Johnson of ABC News, and Bob Arnot of NBC News. And a variety of professional-development programs at colleges and universities, many of them with the financial support of foundations such as the John S. and James L. Knight Foundation, offer fellowships to journalists so they can spend time learning about science and technology.

But although science and medical journalists are better educated, or as well educated, as their peers and journalistic supervisors, they remain less educated than the scientists and physicians whom they cover. This difference in education and training contributes to a gap in status between science and medical journalists and their principal sources. "In no other set of relationships within popular communication, I would argue, do you get more pronounced status differences than between scientists and journalists," observes Sharon Dunwoody.[28]

Scientific Communication

Communication is integral to the process of scientific inquiry. The popular notion of a lone scientist secretly toiling long hours in a laboratory and then striking upon a big discovery, which the scientist announces to the world with great fanfare, is a romanticized fiction. Scientists communicate with each other continually throughout the research process, from the formulation of research questions through applications for grants, the development of methodologies, the interpretation of data, and the review of articles and other research reports. An individual scientist's record of written research communications helps determine whether that scientist will receive tenure at a university or receive research grants. Indeed, molecular biologist Francis Crick, codiscoverer of the structure of DNA, has said that "communication is the essence of science."[29]

To a large degree, popular-science and medical journalism is based on journalists monitoring these communications among researchers, such as scientific conferences and scholarly journals. The journalists select newsworthy communications, translate them into a vocabulary that is understandable to the public, and then add context and background to help readers and viewers make sense of the material.

Of course, science and medical journalists are not limited to monitoring scientific communications. They identify news through other means as well, such as covering the government's handling of scientific and medical issues, congressional hearings, federal financing of research, and the Food and Drug Administration's scrutiny of new pharmaceuticals and medical devices. They may also cover science-related events—space-shuttle missions or advances in medicine such as the implantation of an artificial human heart. They may visit laboratories and see research being conducted in the field, perhaps as far away as the South Pole. (The National Aeronautics and Space Administration [NASA] at one point planned to select a journalist to fly in the space shuttle.) But if communication is the essence of science, scientific communication is the essence of science journalism in that it defines much of how journalists work and their relationship to the scientific establishment.

Scholars often communicate tentative early research findings among themselves informally, through such means as telephone conversations, e-mail exchanges, departmental seminars, and face-to-face meetings. In fact, scholars are understood to operate within an "invisible college" of researchers in their field, at their institution and elsewhere, who are linked by a common interest in a certain subject area. Members of the invisible college exchange insights and data and challenge and critique each other's work. One scholar suggests that new online technologies such as e-mail discussion groups may foster the creation of "electronic invisible colleges."[30]

In general, journalists have tapped into these informal discussions only sporadically. One important reason was a lack of access: scientists averse to popular publicity typically would not include journalists in their "invisible colleges." Another reason was the insistence, particularly of seasoned science reporters, on reporting only findings that had passed rigorous peer review. Indeed, evoking Watergate-era standards of the use of confirmatory sources, one recent critique of science journalism suggests that if a journalist seeks to write about research that has not yet been peer-reviewed, the journalist should seek a "second source" to verify the information.[31]

In the pre-Internet era, a common method of informal communication among scholars—sometimes occurring before a conference presentation, and sometimes after a conference presentation—was distribution of a "preprint," or draft of an article that has not yet been submitted for publication. Researchers distributed preprints to colleagues to get their comments and suggestions about the paper, to allow fine-tuning before it was submitted for publication. William Garvey, in a study of scientific communication in the late 1970s, found that scientists distributed preprints "on a highly personal and selective basis, usually to persons known to be working in the same area."[32]

With the advent of the Web, in many disciplines paper-based preprints have been supplanted by online archives of papers submitted by scholars. The pioneer in this effort was Paul Ginsparg's high-energy-physics archive, which started with 160 users, culled from lists of electronic-mail addresses. More recently, online archives of unpublished papers have developed in many disciplines, such as economics, psychology, and political science. One recent survey found wide variation in the use of electronic preprint archives. More than half of the scientists surveyed in physics or astronomy reported using electronic archives, for example, whereas 4 percent of biological-sciences researchers and no chemists reported using such archives.[33]

Although scholars are the intended audience of these preprint servers, there generally is nothing to stop reporters from retrieving copies of the papers and reporting on them as news. Indeed, the American Physical Society issues "Physics Tip Sheets," press releases that highlight noteworthy physics research, including articles posted on physics preprint servers. However, even with such assistance in locating newsworthy preprints, few science journalists cover preprints, with some notable exceptions such as the weekly magazine *Science News*. In 1998, for example, physicists posted a preprint about an experiment that found that subatomic particles known as neutrinos have mass, but journalists were either unaware or ignored the preprint, even though it was cited in a paper in *Science*. But a week after the *Science* paper was published, a group of researchers issued a press release touting the same results, which generated extensive news coverage. "It is an interesting commentary on science journalism that this press release generated front page headlines," one astrophysicist has observed.[34]

After informal exchanges, conferences often represent the next step in communicating research by one scholar to colleagues. This step is more formal than a casual exchange with a colleague but less formal than a published journal article. At a conference, a scholar can present new findings, get reactions from other specialists in the field, and forge working relationships with peers. Peer review plays a minor role in some conferences, at which virtually any scholar is able to present a paper. At other conferences, peer review does play a role, with potential speakers submitting an abstract of their planned talks and peer reviewers deciding whether to approve the speaker. At still other conferences, potential presenters must submit completed papers for peer review. Of course, the scholar's presentation—along with interchanges with the audience and feedback that the author receives after the presentation—is also a form of peer review.

Conferences are part of the scientific-communication process that science and medical journalists do monitor and report upon. One reason is the semipublic nature of these conferences: they are generally open to accredited journalists, so conferences do not present the access problems posed by attempts to tap into more informal interactions among members of the invisible college. Moreover, because conference presentations are at least minimally peer-reviewed, journalists apparently feel more confident that they are covering valid science.

Journalists' confidence in the validity of research presented at scientific conferences may be misplaced. One recent study found that 26 percent of studies presented between 1989 and 1998 at the annual meeting of the American Society of Clinical Oncology (ASCO)—a conference that draws much media attention—were never subsequently published in peer-reviewed scientific journals; reasons frequently cited by the researchers included insufficient time or funds. Fewer than half of the research results presented at the annual meeting of the Pediatric Academic Society were published in a peer-reviewed journal within five years, a finding that "should cause practitioners, conference organizers, and the press to pause for consideration. Just how much credence should be given to early research findings is unclear." A third study found that about one-fourth of papers at five major biomedical conferences that were covered by the popular media never resulted in published journal articles. Many of the conference papers that were covered by the popular media were of lesser quality, this study found. "The current press coverage of scientific meetings may be characterized as 'too much, too soon,'" the study's authors concluded.[35]

Although some conferences do receive a high degree of media attention, it does not necessarily follow that science journalists spend a large proportion of their time covering such conferences. As the budgets of news organizations have tightened in recent years, travel to scientific conferences appears to have declined. Indeed, one scholar portrays science journalists as tied to their desks—and telephones. "Because science reporters today seldom go to scientific meetings or press conferences, they rarely meet scientists and researchers face-to-face. Thus they depend heavily on the telephone for contacting and connecting with their sources. The telephone is their main link to experts; journalists may develop long-term relationships with sources without ever meeting them."[36]

Scholars in some fields publish their research in monographs, or books, whereas scholars in other fields publish books rarely, if at all. Scholarly publishers do subject books to a form of peer review, by asking outside reviewers to critique a book's text. However, science and medical journalists rely on books as a source of news stories rather rarely. New books do not appear with the predictability of a journal's issues, making them less reliable as a source of news. On occasion, however, scholarly books do receive journalistic coverage, such as a seven hundred–page study of Americans' sexual practices, published in 1994. The book was distributed to journalists who signed an agreement to abide by an embargo, but the arrangement fell apart after the embargo time was revised to accommodate ABC News's *20/20* program and several newspapers decided to publish their stories before the embargo.[37]

However, for many scholars, the most formal mechanism for communicating their research findings to other scholars is through scholarly journals. In general, a scholarly journal is a periodical containing scholarly articles that have been vetted through a peer-review process. Each scholarly field has a hierarchy of journals, with some considered elite journals—such as the *New*

England Journal of Medicine and the *Journal of the American Medical Association* in medical research—and others seen as less significant. In addition, there are a few multidisciplinary journals, such as *Nature* and *Science,* that are generally viewed as publishing the most significant research from many fields of scholarly inquiry. The most prestigious journals are generally ones that have been in existence for a long time, which generally also means that they are published by scholarly societies.[38]

There also is a variety of journal types. Research journals publish formal papers describing scholars' research, letters journals publisher shorter accounts, proceedings journals publish papers that were presented at conferences, review journals are annual volumes that examine a field in depth, and abstracts journals contain only abstracts of research papers.[39]

Some journals are published by not-for-profit scholarly societies or university presses; many others are published by for-profit publishing companies, either independently or in association with a scholarly society. *Science,* for example, is published by the American Association for the Advancement of Science (AAAS), a nonprofit organization. The *New England Journal of Medicine* is published by the Massachusetts Medical Society, and the *Journal of the American Medical Association* is published by the American Medical Association (AMA). *Nature* is published by Macmillan Publishers, a British company.

Journal publishers derive revenue from a variety of sources, including subscriptions (to both individuals and institutions), advertising, and the sale of reprints. Marketing thus is an important consideration for many journal publishers, and media coverage can be a useful form of marketing.[40] A journal can encourage this coverage by distributing a press release that highlights journal articles that may be of interest to the media, according to a recent manual on journal publishing. "There should normally be an embargo on publishing so that the layman does not have information before subscribers get their copies," the manual advises. Journal editors believe that press coverage can benefit their publications by attracting both subscribers and cutting-edge papers. The journal-publishing manual instructs: "One of the truest sayings in the business is that good articles sell subscriptions."[41]

The embargo serves this end. "To be honest, *Science* and *Nature* and other general science journals compete with each other for papers partly through the amount of publicity that they give authors who publish in them," according to Laura Garwin, the former North American editor of *Nature.* "It's a marketing tool," added Monica Bradford, managing editor of *Science.*[42]

*　*　*

Although for-profit publishers by definition seek to profit from the journals that they produce, revenue considerations are not irrelevant for scientific societies. Even for them, journals can be an important source of revenue, and society memberships as well, if a subscription to a society's journal is a membership benefit.[43] The American Association for the Advancement of Science, for example, reported $23.4 million in advertising revenue from *Science* in 2004.[44]

Scholars publish their research in journals for many reasons: to advance their discipline, to claim the intellectual priority on discovering a phenomenon, to contribute to their academic credentials so they can receive tenure or promotion, and so on. In many cases, scholars consequently desire their paper to be published in a journal with high prestige among their professional colleagues, both to lend importance to the paper and to ensure its wide dissemination through the profession.[45] For example, one cancer researcher has said: "I don't think there is any question about the fact that we send it to *The New England Journal of Medicine* because it is a classy journal and [if] you publish it in *Science* and the *New England Journal of Medicine* it means something in terms of the advancement of your career."[46] As a result, high-prestige journals receive many more submissions than they can possibly publish, and competition among authors can be stiff. In 2003, for example, the *Journal of the American Medical Association* accepted 9 percent of the manuscripts that were submitted to it. When attempting to publish a specific paper, an author may initially submit it to a prestigious journal; if that journal rejects the paper, the author may submit it (revised or not) to another lesser light. Some papers are submitted to several journals before they are published.[47]

Many scholars believe that this winnowing process results in the better, or more significant, papers being published in elite journals, with lesser papers relegated to less elite journals. However, because of the well-documented vagaries of peer review, such assumptions appear to be at least open to challenge: elite journals have published papers that have subsequently turned out to have been in error or fabricated, whereas less elite journals at least occasionally publish work of significance.

For some researchers, having an article published in a scholarly journal is not sufficient; they want the article to be the journal's cover story. These researchers go to great lengths to get onto the cover, by supplying photographs or artwork that they think will catch the interest of the journal's editors. For example, one team of researchers whose paper in *Cell* dealt with mammals' sensory receptors for sweetness had a photographer spend seven hours creating an image of two mice sniffing a luscious pastry, which the journal ran on the cover. "A cover appearance can indeed be a big deal, especially for young scientists trying to make a name in their field," one magazine noted. "Although few believe that cover stories can make or break a career, covers nevertheless look good on a curriculum vitae and add heft to presentations at important scientific meetings."[48]

Science and medical journalists heavily rely on these journals as fonts of news. One scholar goes so far as to maintain that "the 'beat' of science fundamentally consists of the elite peer-reviewed journals." One content analysis of science stories in eight U.S. newspapers between 1991 and 1996 found that research reports, such as journal articles, were indicated as the source of 3.3 percent of stories. But some of the newspapers relied on research reports much more than others: research reports were cited as the source for 75 percent of the science articles published by the *Chicago Tribune,* 52 percent of the coverage by the *Christian Science Monitor,* and 50 percent of the articles in the *New York Times.*[49] The content analysis described in the Appendix of twenty-five U.S. daily newspapers found that they published a total of 2,843 breaking-news articles about the *Journal of the American Medical Association, Nature,* the *New England Journal of Medicine,* or *Science* between June 1997 and May 1998.

Science and medical journalists outside the United States also rely on journals in their reporting. In science coverage in Canadian papers, journals were named as a source in 6 percent of stories. An analysis of coverage in five British newspapers found that journals served as the source of 15 percent of the newspapers' stories, with scientific meetings accounting for another 9 percent; more than half of the stories dealt with health and medicine. Over a fifty-year period, journals were the source of 6 percent of science stories in Italian daily newspapers.[50]

Although some of these percentages may seem small, a handful of journals account for most of the journal-related news coverage, so these few journals represent major sources of news for science and medical journalists. One analysis, of genetics-research coverage from 1995 to 2001 by newspapers in the United States, Canada, Britain, and Australia, found that 31 percent of the newspaper articles were based on *Science,* 19 percent cited *Nature,* 16 percent were based on *Nature Genetics* (published by the same company as *Nature,* and with an embargo), and 16 percent were based on *Cell,* which also offers embargoed access to journalists.[51]

The dominance of the elite journals was also documented by a recent analysis of eight and a half years of medical coverage by one newspaper, the *Minneapolis Star Tribune*. That analysis found that ten scholarly journals served as sources for 52.3 percent of the newspaper's articles that cited a scholarly journal. The ten journals—led, in order of decreasing frequency, by the *New England Journal of Medicine*, the *Journal of the American Medical Association*, *Science*, and *Nature*—all offer embargoed access to journalists.[52]

This heavy reliance on a small circle of journals, which are dominated by research conducted by scholars in academe, may be one of the causes for the fact that media coverage of science tends to focus on scientific research conducted by researchers at colleges and universities and tends to ignore research conducted in industry and government, whose researchers often are not under the same pressure to publish in journals as are academic researchers. As one scholar has described this pattern of coverage:

> It is the science of the academic community that has come to serve as the exemplar for popular science coverage. That is, although there are manifest differences between the means and ends of science conducted in military, industrial, and academic settings, it is university science that has come to dominate press attention, and that has commandeered the terms in which science is to be understood and depicted within the overall project of science journalism.[53]

Journalists monitor only a very small fraction of the scholarly journals that are published. Many journalists believe that the most newsworthy research is published in a handful of elite journals, on which the journalists focus their attention. Another major reason is the fact that scholarly journals are legion. Elsevier Science, a division of the publishing giant Reed-Elsevier and the largest publisher of academic journals, alone publishes more than five thousand titles. The American Chemical Society publishes much fewer journals—twenty-seven in all—but even they collectively include more than eighteen thousand peer-reviewed articles in a year.[54] More recently, the Internet has become home to a burgeoning number of scholarly journals that are published only online, with no printed version. Although printed journals are often far too costly for media organizations to afford subscriptions, some online journals are free to all, including journalists. These include a group of life-sciences journals in a federal archive called PubMed Central (http://www.pubmedcentral.gov). The Public Library of Science (http://www.plos.org), a nonprofit group, publishes two journals, *PLoS Biology* and *PLoS Medicine*, which are free online and for which in-print subscriptions are available at low cost.

By comparison to this huge population of journals, even the most hard-working science or medical journalist regularly follows only a comparatively few journals. A medical journalist for a Canadian daily newspaper reports that he "scans about 20–25 journals a week." A guidebook published by the National Association of Science Writers, emphasizing journalists' workload, cites one unidentified editor as reporting that he reads "58 different magazines a month."[55]

One reason that journalists rely so heavily on articles that are published in a handful of elite journals may be that they believe that since an article has passed the journal's peer-review process, it has been certified as a valid scientific finding. "If a study does not appear in a refereed scientific journal it simply isn't science. It may NOT be good science if it does, but it certainly isn't if it doesn't," says Peter Gorner, science reporter for the *Chicago Tribune.* Indeed, a recent critique of science journalism suggested that a journalist could appropriately report on any article that has been peer-reviewed. "Peer review and formal publication . . . give journalists a comfort zone by guaranteeing that the work presented is accurate to the best of the experts' knowledge."[56]

Some scientists believe that journalists do not place enough faith in peer review. For example, Robert McCall argues that journalists are more concerned with quoting a scientist accurately than in determining whether the scientist is correct (or, as McCall puts it, "whether the scientist has faithfully reflected current scientific knowledge").[57]

But other scholars are not so sanguine about the effectiveness of peer review, identifying many flaws. Statistical errors often are missed by reviewers, calling into question the ability of peer review to sort the wheat from the chaff. Four scholars recently reviewed twenty-one studies that examined various facets of peer review and concluded: "There is little empirical evidence to support the use of editorial peer-review as a mechanism to ensure quality of biomedical research, despite its widespread use and costs."[58]

Others suggest that peer review suppresses innovative ideas. One researcher has cataloged more than a dozen important research results in medicine that challenged the status quo that were rejected by major journals. This concern extends beyond biomedicine: Frank J. Tipler, a professor of mathematical physics at Tulane University, says that peer review has often stymied the development of new scientific ideas because the "peers" who conduct the reviews are often less capable than the authors who have submitted papers: "We have pygmies standing in judgment on giants."[59]

Journalists' faith in peer review is not absolute. David Perlman, the former science editor of the *San Francisco Chronicle,* says he often feels the need to question the validity of peer-reviewed research: "Who's behind it? Is somebody trying to peddle something? Who's the researcher? You develop some kind of instinct for balderdash."[60]

However, journalists' vetting of peer review also is far from perfect. A case in point was media coverage of an article published in the April 1, 1998, issue of the *Journal of the American Medical Association,* which claimed to disprove the effectiveness of "healing touch," or using one's hands to manipulate a "human energy field" around a patient's body. The AMA included the article as the first entry in its weekly embargoed press release for journalists, and the media responded with enthusiastic coverage.[61] One factor motivating the extensive coverage undoubtedly was the fact that the experiment had been done by a nine-year-old for her fourth-grade science fair; one of her coauthors was her mother. All three of the national broadcast networks covered the journal article; only two other journal articles out of the 2,655 articles published by the *Journal of the American Medical Association, Nature,* the *New England Journal of Medicine,* or *Science* between June 1997 and May 1998 received coverage by all three networks. None of the three network reports on the healing-touch study mentioned any limitations or criticisms of it.

Newspapers also covered the story widely: seventeen out of a group of twenty-five U.S. daily newspapers covered the healing-touch paper; only eleven articles published in any of the four leading journals during the twelve months ending May 1998 received coverage by more newspapers. And the coverage was upbeat in tone. A page-one story in the *New York Times,* for example, said that the paper had "thrown the field into tumult" with a "devilishly simple" experiment. A front-page story in the *Los Angeles Times* noted criticism of the study only briefly and reported: "Using little more than a towel and a piece of cardboard, a 9–year-old girl conducted a 'brilliant' study debunking therapeutic touch. . . . Along the way, Emily Rosa, now 11, has apparently become the youngest researcher to publish a scientific paper in the prestigious *Journal of the American Medical Association.*" The Associated Press's story, which several of the newspapers carried, limited criticism of the paper to one of the founders of the therapeutic-touch movement and immediately followed that with a comment from the journal's editor defending the study.[62]

Despite the approving media coverage, the study was received with much more skepticism by scientists. One letter to the journal called the study "simpleminded, methodologically flawed, and irrelevant." Others complained that the two researchers were not unbiased, because the mother was a nurse for an organization that seeks to discredit healing touch. Clearly, in at least this situation, the science journalists' efforts to double-check the peer-review process failed to unearth criticism of the paper.[63]

The Ingelfinger Rule

The scientific establishment uses two tools to regulate interactions between researchers and the press. One, commonly known as the "Ingelfinger Rule," has the effect of pressuring scientists to be chary of discussing their research with journalists before publishing the research in a journal. The other, called an embargo, is a more direct restriction on journalists themselves, in which journalists get advance access to journal articles under the condition that they refrain from publishing news about those articles until a preset time.

The Ingelfinger Rule refers to a policy that states that a given journal will not publish a scientific paper that has been already disseminated, particularly through the popular press. The rule derives its name from Franz Ingelfinger, who forcefully enunciated and enforced the policy while he was editor of the *New England Journal of Medicine* from 1967 to 1977. Although the rule actually predates Ingelfinger, as Chapter 2 will relate, he first articulated his rule in 1969 after a medical researcher submitted to the journal a paper whose essence—"including its one and only illustration"—had already been published in the *Medical World Tribune*. The journal had already stated that it would publish only papers that had not been published or submitted for publication elsewhere. But the incident spurred Ingelfinger to issue a more specific statement that he would publish only papers that had been "neither published nor submitted elsewhere (including news media and controlled-circulation publications)." Ingelfinger exempted news media coverage of talks by medical researchers at scientific meetings but clearly signaled researchers that they should use extreme caution in conducting interviews with reporters that would expand on the often sketchy information contained in such talks: Ingelfinger declared that he would refuse to publish a paper "if the speaker makes illustrations available to the interviewer, or if the published interview covers practically all the principal points contained in a subsequently submitted manuscript."[64]

Ingelfinger appeared to justify the rule purely on competitive grounds—he wanted his journal to publish articles that had not appeared elsewhere. But in the decades since, other medical journal editors have developed another reason for the rule: the need to discourage dissemination of medical research that has not been validated by a peer-review process. In this view, the Ingelfinger Rule prevents media coverage of research that cannot pass peer review. For example, two editors of the journal *Neurosurgery* recently argued: "Substantive, meaningful, and accurate information must be presented so that the public can trust what they read and so that doctors are not besieged with questions irrelevant to patient care."[65]

Ingelfinger and his successors have articulated several exemptions to the rule, including public release of data by public health authorities, disclosure to congressional committees or government regulatory bodies, and special cases where immediate public release is appropriate, such as results of clinical studies of AIDS therapies.[66] There is evidence that the rule nevertheless slows the diffusion of medical research to physicians and, consequently, the availability to patients of new or improved therapies. One study found a rapid increase in the use of a surgical procedure known as carotid endarterectomy, which is intended to prevent strokes, after clinical alerts describing favorable results of two trials were released to the public prior to the results being published in peer-reviewed journals.[67]

The rule enjoys wide support among journal editors. Out of a group of 80 major journal publishers, almost three-fourths follow the Ingelfinger Rule. More than 500 journals now adhere to a set of voluntary principles, called the Uniform Requirements for Manuscripts Submitted to Biomedical Journals, which includes a proscription similar to the Ingelfinger Rule. Many major nonbiomedical journals, such as *Science* and *Nature,* also refuse to publish scientific papers that have been publicized in the popular press. And in a survey of editors of 269 leading medical journals published in the United States and Canada, Michael Wilkes and Richard Kravitz presented respondents with a summary of the Ingelfinger Rule—though not identified as such. Asked for their views, 77 percent of respondents said they agreed with the rule. (Interestingly, 84 percent of the respondents also agreed with a further statement that once a study is published, scientists should be required to speak with the press. Wilkes and Kravitz do not explore the reasons for editors' interest in compelling authors to interact with the press, but it is possible that the editors are interested in the publicity that such interactions would provide for their own journals.)[68]

However, critics of the Ingelfinger Rule argue that it is based on the flawed premise that peer-reviewed studies are more scientifically valid than studies that have not yet been peer-reviewed, that much medical and scientific research is conducted with public funds and therefore should be open to public scrutiny even before peer review, that the rule deters scientists from providing public officials with early access to scientific data that may be crucial for informed policy decisions, and that the rule fosters unwarranted distrust of reporters by researchers and thus impedes effective interaction between them.[69] Lawrence K. Altman, a physician and medical correspondent for the *New York Times,* argues that proponents have not documented any benefits from the Ingelfinger Rule: "Independent analysis is needed to document contentions that the Ingelfinger rule improves and assures the quality of what journals publish. Unless and until supporters provide such evidence, the Ingelfinger rule should be dropped."[70]

The Ingelfinger Rule's restrictions on prepublication publicity often are confusing to scientists, who worry that they may jeopardize their chances for publishing research in a prestigious journal by even a minor misstep with the media. The journals promote a cautious approach. *Science,* for example, offers the following advice for scientists who wish to make a presentation at a scientific conference of material that they have submitted for publication in *Science:* "Comments to press reporters attending your scheduled session at a professional meeting should be limited to clarifying the specifics of your presentation. In such situations, we ask that you do not expand beyond the content of your talk or give copies of the paper, data, overheads, or slides to reporters." *Nature* also warns scientists against cooperating with journalists: "Communicate with other researchers as much as you wish, but do not encourage premature publication by discussion with the press (beyond a formal presentation, if at a conference)." If the media do cover a paper that has been submitted to the journal, the editors warn: "*Nature* will assess the extent to which authors have solicited this interest or cooperated with journalists. If, in the judgement of the *Nature* editors, *Nature*'s embargo policy has been broken, the submitted paper may be rejected, even if it is technically 'in press.'"[71]

In the face of such warnings, it is unsurprising that scientists who speak at scientific meetings are often uncooperative with journalists who cover the meetings; the journalists often find themselves confused by the fast-paced and cryptic presentations devised by researchers to disclose their research findings without breaking the Ingelfinger Rule.[72] As a consequence, since the rise of the Ingelfinger Rule and its variants, journalists have found scientific and medical meetings to be a less fruitful source of news stories. One health-care journalist says of the Ingelfinger Rule: "Almost every health or medical journalist can recount cases of researchers too terrified to share even basic clarifications or explanations of their work with reporters, for fear that their papers would be banished from consideration."[73] This decline in the news value of scientific meetings may correspond to the rise in journalistic interest in peer-reviewed journals.

One place where the Ingelfinger Rule holds less sway is at the online journals published by the Public Library of Science. The nonprofit group asks authors of journal articles to "not contact the media or respond to such contact unless an article has been accepted for publication and an embargo date has been established." But the group does not threaten to pull publication of an article if it is covered by the popular press: "If a journalist has covered a piece of work ahead of publication, this will not affect consideration of the work for publication."[74]

Embargoes

Many major scholarly journals distribute advance information from each issue—perhaps copies of selected articles or even the full issue—to science journalists before their readers or the general public, on condition that the journalists do not disseminate news coverage of the articles until a predetermined time that is common to all the journalists who participate. This arrangement is known as an embargo. The material is said to be *embargoed* until the release time.

For medical journal editors, a cardinal virtue of the embargo is that its timing is arranged so that news coverage of a journal article does not occur until the issue of the journal containing that article has been delivered to physicians. This, editors argue, gives physicians an opportunity to read and digest a journal article before patients, who have heard or read about the article through the popular media, start calling with questions about it. As the *Journal of the American Medical Association* describes the situation: "Advance copies of JAMA are mailed to physicians beginning approximately 1 week before the Wednesday print cover date. This allows physicians to receive their copy of The Journal and have access to the upcoming articles before any news coverage occurs so they will be prepared if patients contact them about news reports based on material in that issue." Physicians believe that if medical news surfaces before the peer-reviewed journal article is available for them to consult, they and their patients are at a disadvantage: "There is nothing more confusing or troubling to a patient than bringing information to a physician asking for guidance, care, counseling, support, advice, when the physician has not had the opportunity to at least assimilate even the briefest synopsis of the information."[75]

Specifics of the embargo vary from journal to journal. Many journal publishers distribute lengthy press releases each week that describe articles in the journal's forthcoming issue. Journalists then may request that the full text of specific articles be sent to them. Some universities, foundations, and companies also distribute embargoed press releases about journal articles with which they are associated—for example, the university at which an article's author works might produce a press release highlighting the university's connection with the work.

Over the past decade, the development of online technology has transformed these exchanges. Before the popularity of the Internet, the press releases were distributed by first-class mail, and reporters would telephone requests for articles, which would be sent by fax, courier, or overnight delivery service. More recently, though, journal publishers have embraced the World Wide Web as a delivery mechanism for both the press releases and the embargoed texts of articles.

For example, the American Association for the Advancement of Science, which publishes *Science,* makes that journal's weekly embargoed press release and embargoed material from several other journals—including journals of the American Chemical Society, the *Astrophysical Journal, BMJ,* the *Journal of the American Medical Association,* the *Journal of the National Cancer Institute,* the *Lancet, Proceedings of the National Academy of Sciences,* and nine journals published by Cell Press (*Cell, Neuron,* and others)—available to approved journalists on its EurekAlert! Web site (http://www.eurekalert .org), established in May 1996. Although a section of the site is open to the public, embargoed material is limited to a password-protected section; each journal publisher that participates in EurekAlert! can decide whether to grant a given journalist access to that journal's embargoed material. EurekAlert! also accepts press releases—both embargoed and open to the public—from universities, research centers, and companies.

In the case of *Science,* the embargoed press release on EurekAlert! includes links to the full text of the articles, in Adobe portable document format, so journalists can download any or all articles and read them on their own computers without the cost and time delay of relying on overnight couriers. Alternatively, journalists can request that the texts of *Science* articles be faxed to them. The press release lists all the articles in the upcoming issue of *Science* and contains nontechnical explanations of most of them. The press release also lists the institutional affiliations of the authors of the papers to help journalists who are looking for local researchers who have published papers. *Science*'s embargo expires at 2 P.M. eastern time on Thursday, a time that permits coverage by late-evening television in Europe and early-evening local news broadcasts in the United States.

Science's embargo is binding only on those journalists who agree to participate in the embargo. But the American Association for the Advancement of Science does try to extend the embargo to cover nonparticipating reporters; the association forbids scientists who have a paper awaiting publication in *Science* from participating in any news conference about their paper earlier than 1 P.M. on the day of the embargo—that is, more than an hour before the embargo lifts. The scientists are also permitted to grant interviews only to journalists who have agreed to the terms of the embargo, and only in the week before publication. These restrictions leave scientists with few options for cooperating with reporters who learn about research through independent channels; the implied threat is that *Science* may decide not to publish the paper of a scientist who breaks its rules about dealing only with journalists who agree to operate by the embargo rules.

EurekAlert! does not charge journalists, but it does charge journal publishers and groups that post press releases on it. These organizations pay the fee because EurekAlert! has proven very popular among science journalists. In addition to the journal publishers noted above, several government agencies—such as the National Institutes of Health (NIH) and the U.S. Energy Department—post material on EurekAlert! and the site's high profile among journalists makes it a popular location for nonprofit groups to disseminate material to journalists. Indeed, by spring 2001, a cumulative total of more than twenty thousand press releases had been posted on EurekAlert! As of April 2003, more than forty-four hundred journalists were registered to use the site, half of them outside the United States.[76]

The popularity of EurekAlert! outside the United States—as well as the general predominance of U.S. journals in science news—prompted the development of AlphaGalileo (http://www.alphagalileo.org), a similar Web site for disseminating news about scientific research conducted in Europe. "Alpha-Galileo grew out of concerns shared by many researchers across Europe, but particularly in the UK that our newspapers and television carried a preponderance of news of American science achievements and that they rarely covered European developments as extensively. There was a growing perception amongst young people that to do exciting science you had to move to the USA," the site explains. AlphaGalileo was started by European science agencies and initially operated by the British Association for the

Advancement of Science, but in April 2003 it was put under the aegis of an independent foundation. By summer 2003, about four thousand journalists from eighty countries had registered to use the site.[77]

Nature has also taken advantage of the Internet for disseminating embargoed material. Macmillan Publishers distributes embargoed press release for *Nature* and its sister journals (including *Nature Biotechnology, Nature Cell Biology, Nature Genetics, Nature Immunology, Nature Materials, Nature Medicine, Nature Neuroscience,* and *Nature Structural Biology*) through e-mail. Journalists can also request a password for http://press.nature.com, which contains both the press releases and the texts of journal articles for downloading. *Nature*'s press releases are more selective than *Science*'s, generally describing only a few of the articles in a given issue, but prominently providing geographic information for reporters seeking local angles. *Nature*'s embargo time is 6 P.M. London time, or 1 P.M. U.S. East Coast standard time.

In addition to using EurekAlert! the *Journal of the American Medical Association* has its own password-protected Web site (http://www.jamamedia .org) from which approved journalists can download embargoed texts of upcoming articles from that journal and the association's *Archives* journals (*Archives of Dermatology, Archives of Facial Plastic Surgery, Archives of General Psychiatry, Archives of Internal Medicine, Archives of Neurology, Archives of Ophthalmology, Archives of Otolaryngology—Head & Neck Surgery, Archives of Pediatrics & Adolescent Medicine,* and *Archives of Surgery*). The *Journal of the American Medical Association* distributes advance copies of the entire journal, a press release describing several of the articles and editorials in the issue, and video news releases on selected journal articles. The embargo time is 3 P.M. central time on the day before the cover date on the journal—which translates to 3 P.M. Tuesday for the *Journal of the American Medical Association* and 3 P.M. Monday for the *Archives* journals. The journal is mailed to physicians in time for them to have several days to read the journal's articles before the embargo time, so that they will be able to answer questions from patients that are prompted by news coverage.[78]

The embargo system at the *New England Journal of Medicine* permits reporters to subscribe to the journal by first-class mail, as opposed to the second-class delivery provided to regular subscribers. This enables the journalists to receive copies of the journal several days earlier than regular subscribers, but the journalists agree not to disseminate news stories until the evening before the journal's cover date, a release time that provides them with several days to research their stories. This journal produces no press releases, although other institutions such as universities and pharmaceutical companies do produce embargoed press releases tied to research published in the journal. One of its editors suggests that editorials in the journal that comment on research in the journal do help science journalists identify the most newsworthy reports in an issue.[79] Its embargo time is 5 P.M. eastern time on Wednesdays, which is set to coincide with evening local news broadcasts.

In Britain, advance copies of the medical journals the *Lancet* and *BMJ* usually reach journalists on Thursday, along with press releases from each journal indicating potentially newsworthy articles. The release time for the embargoes on both these journals is Friday morning.[80]

The *Proceedings of the National Academy of Sciences of the United States of America,* which is published on the second and fourth Tuesdays of every month, puts a different twist on the embargo by publishing its articles on its Web site (http://www.pnas.org) before publishing them in print. The journal, known to scientists and science journalists as *PNAS,* distributes an embargoed tip sheet for journalists each Thursday. The tip sheet lists several articles that the journal plans to publish on its Web site sometime during the coming week; like the press releases produced by other journals, these press releases contain a lay-language description of many of the papers and the locations of the authors to help journalists in finding local angles to stories. However, the *PNAS* tip sheet does not list the specific day when each article will appear online. On Monday evening, the embargo lifts, and journalists are allowed to report on any of the papers that are scheduled for online publication during that week.

Science journalists have access to other embargoed information as well. One of the most frequently used is the National Academy of Sciences, an independent body that examines science-policy questions for the federal government, which often provides journalists with embargoed copies of their studies one or more days before the scheduled release time. The academy does not require journalists to sign an embargo agreement, but it apparently is understood that journalists who violate the embargo may be cut off from future embargoed access to the academy's reports. Commercial public-relations services, such as Newswise (http://www.newswise.com), also provide embargoed materials to journalists on behalf of universities or journal publishers.

Despite the importance that journalists and the scientific establishment attaches to the embargo arrangement, little scholarly attention has been devoted to it. One study recently examined the processes by which seven major medical journals—*Annals of Internal Medicine, BMJ, Circulation,* the *Journal of the American Medical Association,* the *Journal of the National Cancer Institute,* the *Lancet,* and *Pediatrics*—handle press releases. The researchers found that, at all seven journals, the journal editors selected which journal articles were sufficiently newsworthy to merit press releases, and the releases themselves were written by communications officers, in some cases without any further involvement from editors at the journal. Another study examined the handling of the press by the American Association for the Advancement of Science regarding the finding of possible fossilized bacteria in a meteorite from Mars; the association's official in charge of the embargo was willing to consider manipulating its terms to better control the press, such as by offering special early access to the *New York Times* in an effort to forestall that paper from discovering the research independently outside of embargoed channels.[81]

Supporters of journal embargoes offer a constellation of justifications for it. Perhaps the most common argument, and one that is advanced by both journalists and journal editors, is that the embargo provides journalists with time to execute more accurate and more complete journalism about science and medicine. Since participating journalists have several days to prepare their stories about an embargoed study, this rationale maintains, the journalists are less likely to make factual errors. "I like journal embargoes," said Rick Weiss, a science reporter at the *Washington Post*. "They do help make science stories more accurate." Likewise, some journalists say, the embargo's added time gives journalists a better chance to contact a range of experts to comment on a paper; in the crush of an unembargoed deadline, journalists at smaller or lesser-known publications might be unable to reach researchers besieged with inquiries from the press around the world. "The more people you can talk to, the better," says John Travis, a reporter for *Science*. "Time equals people."[82]

Promoters of the embargo point to problems that have arisen when embargoed information has been disclosed to the public prematurely, forcing journalists to scramble to catch up. One notorious example was in 1999, when NBC News reported on a National Academy of Sciences study about medical errors the evening before a one-day embargo on the study lifted. Larry Tye, a *Boston Globe* reporter, complained that he had only an hour to report and write his story. "The reader was getting a glossed-over sense of the problem and probably no sense at all of the causes and fixes," Tye said. Another reporter agreed. "When the embargo is lifted . . . at 5 p.m. just before it goes on the air . . . it results in less communication with the public . . . which is the whole point of the embargo," said Paul Recer, then a science reporter at the Associated Press.[83]

Journalists and journal editors also argue that the embargo is a means of treating all science and medical journalists evenhandedly. "The policy . . . provides a level playing field for all media to report on our original findings," Floyd E. Bloom, the former editor of *Science,* wrote in a typical defense of embargoes. This argument appears to play on a fear among journalists that, in the absence of an embargo, researchers or agencies would favor major news organizations by providing them with access to research results before other journalists. However, it seems clear that an unembargoed arrangement—for example, releasing a study to all journalists at the same time, at which point the journalists would be immediately free to report—could also provide equal treatment for all journalists as well.[84]

ABC News's medical editor, Dr. Timothy Johnson, has invoked the equal-treatment defense of embargoes, arguing that an

> important effect of embargoes is to level the playing field—at least in terms of deadlines—for the journalism community, thereby preventing what would otherwise be a bloody scramble to be first, with an inevitable decrease in the quality of reporting and assessment. . . . I fear the elimination of the embargo policy would quickly result in barbarians at the gate of public disclosure, with scientists and journalists hounding one another mercilessly in an effort to be the first to gain public attention. Indeed, I could imagine scenarios in which prominent scientists are staked out by the press at their homes or offices.[85]

Some believe that embargoes provide other benefits to science. Peter Wrobel, managing editor of *Nature,* says that society benefits from the embargo because it increases the news coverage that a journal article receives. Without an embargo, most news organizations would not cover research published in journals, he claims. "By maximizing publicity for good research, it maximizes the news about science that's actually available to the average reader or viewer," Wrobel states. "More of the good has got to be better than less of it." Weiss, the *Washington Post* reporter, says that the embargo arrangement, in which major journals have staked out different days of the week for their releases, also serves to space out science and medical news stories throughout the week. If the stories from all the journals were to happen on the same day, editors would be unwilling to run all of them, he says. "There's only so much appetite for science in the newspaper."[86]

Despite the prominence of embargoes, however, some sources of scientific news do not embargo their articles or provide press releases to journalists. Several major scientific societies, such as the American Geophysical Union (AGU), the Association for Computing Machinery, and the Institute of Electrical and Electronics Engineers (IEEE), eschew embargoes. In some cases, editors or journal publishers may judge that journalistic interest in their journals' content is too low to justify the move. Of course, to the extent that embargoes spur journalistic interest in a journal, this may be a self-fulfilling prophecy. A case in point may be social science. Social scientists generally bemoan what they see as an inappropriately low level of press coverage of their field.[87] However, social science journals have used embargoes as a tool for promoting news coverage much less aggressively than journals in the biomedical and physical sciences.

In isolated cases, an embargo may not be legally permissible. For example, President Clinton's Advisory Committee on Human Radiation Experiments examined the ethics of experiments on the effects of ionizing radiation in humans that the federal government had conducted from 1946 to 1974—experiments often conducted without the consent of the humans who were the experimental subjects. Because the committee was subject to the Federal Advisory Committee Act, its report—and even draft versions of individual chapters of the report—was freely available to the press and public, without any embargo. Stephen Klaidman, director of communications for the committee, later said: "There were certainly times when all of us felt that raw data were written about in the media in a way that inaccurately reflected what ultimately turned out to be the case . . . but I think that on the whole it was overwhelmingly beneficial to operate in the open."[88]

Other publishers do not use embargoes because, they say, it is inappropriate for scholarly societies to restrict the free flow of information. "The information is already available, and we believe that there is no legitimate reason to keep it artificially under wraps," says Harvey Leifert, the public information manager of the American Geophysical Union who is a particularly vocal critic of embargoes. And he maintains that the embargo actually decreases the amount of public information about scientific issues. At his association's scientific meetings, he says, "some scientists have declined to participate in a press conference because they think they might, at some later date, write a paper on the topic and submit it to *Nature* or *Science.* The embargo actually applies only to papers already submitted to those journals, but so pervasive is the fear of punishment (rejection of the paper and, perhaps, of future papers) that scientists censor themselves just to be on the safe side."[89]

The crux of the embargo arrangement is that all participating reporters agree to withhold their coverage until the same predetermined time. Occasionally, however, the embargo fails, when one or more journalists disseminate their news coverage before the predetermined time. In such cases, journalists generally consider the embargo to have been vacated and that they are free to publish their coverage whenever they desire. "On major stories, it's a good idea to get your story done well in advance of the embargo date, so it will be ready to run right away if someone else breaks the embargo," advises Tom Siegfried, former science editor at the *Dallas Morning News*.[90]

The journalist or media organization that is the first to break an embargo can be penalized by the journal publisher by cutting off their embargoed access in the future. The publisher has complete discretion as to what penalty to apply; the American Association for the Advancement of Science, for example, suspends a willful violator's advance access to *Science* for up to six months.[91] But as Chapter 2 will relate, in many cases rescinding a journalist's embargoed access has backfired, because the journalist then is freed to seek out embargoed studies through unofficial channels and is under no obligation to adhere to the embargo release time. Many journalists suspect that journal publishers often overlook embargo violations by major media organizations, but journal publishers say that they treat all violators equally.

When an embargo violation occurs, many journalists will simply proceed to rush their own coverage into print or on the air as soon as possible. Journal publishers often notify participating journalists when an embargo violation has occurred. Indeed, some feel an ethical duty to journalists to do so.[92] However, journal publishers also attempt to reserve for themselves the authority to rule on whether the embargo remains in force despite a violation by someone in the media. For example, in June 2002, the Institute of Physics, in Britain, complained that the wire service Agence France Presse broke the embargo on research published in one of its journals describing development of a mannequin that can perspire. "PLEASE NOTE THAT THE EMBARGO REMAINS IN PLACE," the director of AlphaGalileo wrote in an e-mail message to journalists registered on that site for embargoed material.

And indeed, although journalists commonly understand the embargo arrangement to be lifted if the embargo fails and one or more journalists report the scientific discovery before the agreed time, the actual embargo agreements contain no such provision. For example, EurekAlert!'s description of its embargo policy provides no exemption for embargo failures. "Only the AAAS Office of Public Programs is authorized to make changes to a paper's embargo time," it says.[93] Similarly, *Nature*'s embargoed press releases contain no "out" for embargo violations.

There is little data on the frequency of embargo violations, but they appear to be rare. One analysis of 1988 coverage of the *Journal of the American Medical Association* and the *New England Journal of Medicine* by ten major U.S. newspapers found that 7 percent of the articles appeared to violate the journals' embargo, generally one or two days early. One official at the American Association for the Advancement of Science estimates that embargo violations for *Science* occur, on average, once every two to three months. "Most of the problems occur Thursday afternoons, very near the regular embargo-release time, and are related to human error, such as an accidental, premature posting to a Web site."[94]

In some cases, a journalist may uncover a story outside the embargo system. Under such circumstances, the journalist is generally understood to be free to publish the story at any time. Once such a story is published, other journalists generally consider the embargo to be without force, and even embargo participants consider themselves free to publish their reporting immediately, if they choose to do so. This often happens with the most important stories—such as the first cloning of a sheep, the claimed discovery of fossilized bacteria in a meteorite from Mars, and results of the sequencing of the human genome. Consequently, in the case of major findings, some journalists have taken to preparing their stories far ahead of the embargo time to be prepared to publish quickly if the embargo fails.[95]

The issue of independent reporting is not always clear-cut. This was evident in an embargo incident in July 2002, when the *Journal of the American Medical Association* published a report from the National Heart, Lung, and Blood Institute (NHLBI) with worrisome findings for women who were taking hormones after menopause. The study found that women who used a combination of the two hormones estrogen and progestin after menopause suffered more heart attacks and strokes than women who did not use the hormones. Because the risks of the therapy clearly outweighed the benefits, the clinical trial in which the hormones were tested was stopped two years earlier than planned. The NHLBI, part of the National Institutes of Health, scheduled a press briefing for 9:30 A.M. on Tuesday, July 9, 2002, to discuss the research. The day before the press conference, the medical journal alerted reporters about the study and the press release under an embargo that expired at the start of the press conference—that is, at 9:30 A.M. on Tuesday, July 9—at which time the journal planned to post the text of the journal article and an accompanying editorial.[96]

However, on the evening of Monday, July 8, the *Detroit Free Press* distributed a story on the findings, via a wire service operated by Knight Ridder Newspapers, of which it is a part. "Hormone replacement therapy—medication taken by millions of American women for relief of menopause symptoms and prevention of other serious health problems—should not be taken long-term because it carries more risks than benefits, national researchers will report today," said the version of the story that ran on the morning of July 19 on the paper's front page and on its Web site.[97]

Patricia Anstett's article, disseminated on the Knight Ridder wire, was published in papers in cities such as Albany, N.Y.; Bergen County, N.J.; Columbia, S.C.; Contra Costa, Calif.; Duluth, Minn.; Montreal; Myrtle Beach, Fla.; Orlando, Fla.; St. Paul, Minn.; and Toronto. Other media outlets followed suit to cover the findings before the press conference. Among newspapers running their own coverage of the findings on the morning of July 9 were the *Atlanta Journal-Constitution, Charlotte Observer, Detroit News, Fort Worth Star-Telegram, Houston Chronicle, Raleigh News and Observer, New York Times, Sacramento Bee, San Francisco Chronicle, Seattle Post-Intelligencer,* and *USA Today.* "The result was fairly widespread television, Internet, and newspaper coverage reporting the . . . findings before the NHLBI investigators had announced the findings at the press conference and before physicians (and women in the . . . clinical trials and others) had access to the full article and the accompanying editorial," the journal's editors later complained.[98]

Consequently, the *Journal of the American Medical Association* cut the *Free Press* off from embargoed access. "I consider this breach extremely serious and have instructed our media relations staff to remove the Detroit Free Press, which initially broke the embargo, from our mailing list immediately," the journal's editor in chief, Catherine D. DeAngelis, told reporters in a memo. "As a result, it will not be receiving JAMA news releases, articles or advanced copies of the Journal. In addition, no future interview requests from the reporter . . . will be granted." But reporters at other media organizations forwarded the journal's embargoed materials to the *Free Press* reporter, Pat Anstett, so the ban had little practical effect on her work.[99]

Some journalists felt the *Detroit Free Press* had acted properly. The ombudsman of the *Atlanta Journal-Constitution,* which was apparently spurred by the Detroit story to run its own article in its July 9 issue, characterized that move as a decision to "publish what many people connected to the study were already talking about." By contrast, the *Washington Post* waited until the agreed date of July 10 to run its story. Its ombudsman suggested that his newspaper should have been faster off the mark. And other journalists suggested that the article should have been run sooner if it was based on independent reporting rather than the embargoed announcement by the journal.[100]

Indeed, the reporter and her editors insisted that they had learned of the study from other channels. "The story I wrote for the *Detroit Free Press* was independently reported, using sources with first-hand knowledge of the study and its abrupt end," Anstett said. She maintained that a researcher had been warning her, over a period of months, that hormone-replacement therapy was doing women more harm than good, and finally tipped her off to the end of the study. In a letter to DeAngelis, Anstett's editor said that Anstett "had done several weeks of independent reporting," with sources who talked with her before JAMA distributed its embargoed information.[101]

The Detroit journalists pointed out that the researchers had notified the 16,608 women participating in the study about its premature end more than a month before the journal's embargoed announcement. "We got wind May 31 that the study was in trouble," another editor at the paper said. In their view, delaying the release of the information through the embargo needlessly deprived women of the information that they needed to manage their own health. Some on the inside were not fettered in this way, Anstett noted: DeAngelis—the JAMA editor—decided to stop taking hormone replacements within twenty-four hours of learning of the study's results, long before most women heard about them. "She had the benefit of inside information but they didn't share it with the public for another month," Anstett said.[102]

Editors at the medical journal maintained that the story was a violation of the embargo regardless of how Anstett became aware of the study's results. In a lengthy editorial published in the journal a month after the hormone study appeared, the journal's editors said that Anstett contacted a physician for an interview on Monday, July 8, and mentioned the embargoed press release. The physician granted an interview, but it was embargoed until the time of the press conference the next day. DeAngelis herself called the reporter on Monday evening to advise her that the story would be considered an embargo break if published before the time of the press conference.[103]

This incident underscores the power that the embargo gives journal editors over journalists: the journal editors not only decide what information to provide to journalists and at what time the journalists can disseminate that information to their readers or viewers but also reserve to themselves the power to regulate journalists' interactions with nonembargoed sources. (Note that DeAngelis did not claim that Anstett knew about the hormone-replacement study only because of the embargoed material sent out by the journal; rather, DeAngelis appeared to argue that once Anstett came into possession of the embargoed information, all her subsequent actions were governed by that embargo.) Under the embargo arrangement, journal editors also retain the authority to investigate alleged breaches of the agreement, to determine guilt or innocence, and to impose punishment on the journalist or their media organization or both.

Anstett is not the only journalist to be caught by the journals' ever-widening assertions of power and authority over journalists. In February 2003, *Nature* published a paper that concluded that the oldest human remains so far found in Australia are about twenty thousand years younger than scientists had previously thought. News coverage was embargoed until the morning of February 20 (Australia time). But Bob Beale, a science reporter for the *Bulletin with Newsweek,* an Australian newsweekly, learned about the research independently and published an article about it in the February 19 issue of the magazine. Although *Nature* agreed that Beale had obtained the information independently, it nevertheless suspended his embargoed access because he was aware that it was under a *Nature* embargo. The journal told Beale: "By signing up to our press service, you are bound to abide by our embargoes in a more all-encompassing way (as opposed to just press release by press release). So, as soon as you became aware that the work you were researching was to be published in *Nature,* you should have waited for notification from us as to when its embargo would legitimately lift, or sought this information from us." One science journalist has summarized *Nature*'s argument: "Even though a reporter has established facts on his own, he should wait on the story if at any time before publication *Nature* magazine independently publishes independent information on the same topic." After eight months, *Nature* agreed to reinstate Beale's embargoed access.[104]

Using the Web to distribute embargoed materials to journalists has provided journal publishers with a new tool for policing embargo infractions: unambiguous information on whether a journalist had embargoed access to a specific journal article. In the past, a journalist could plead that he or she had not seen an embargoed article and instead had independently developed a news report about it. But computer-usage logs maintained by Web sites such as EurekAlert! can specify which journalists received access to embargoed materials and when. Indeed, EurekAlert! has warned its journalist users that it can use computer records in exactly that fashion if a journal or research institution that supplied embargoed material to EurekAlert! complains about a journalist's actions: "If we do agree to provide information related to an alleged embargo break, only three factors may potentially be revealed and only to the relevant journal, institution and/or registered user: whether or not the person viewed the document for which the embargo was broken; if so, what time of day he or she viewed it; and whether his or her request for the file, if made, was successfully fulfilled."[105] Neither the *Nature* press site, the embargoed site of the *Journal of the American Medical Association,* nor the European

Union's AlphaGalileo press site contains such a forthright declaration of the possible use of computer logs for tracking embargo violations, but all three systems presumably have capabilities similar to those of EurekAlert!

Shaping the Content of Science Stories

To understand how embargoes shape the coverage of science by journalists, we must first examine, in general, how journalists and sources interact to construct news. Decades of research by communication scholars have produced two insights about this process that are particularly relevant to embargoes: journalists generally try to make sure that their work is similar to their competitors', not different, and "information subsidies" influence what journalists cover and how they cover it. We will examine each of these insights in greater detail and then incorporate them into a model that attempts to explain the nature of news coverage of scientific journals.

The content of news media—what is covered, and how—is strongly influenced by competition and cooperation among individual journalists and journalistic organizations. Although journalistic folklore suggests that journalism is relentlessly competitive, cooperative behavior is far more common than usually acknowledged. Competition often occurs under sharp limits set with the concurrence of journalists. "A news organization would usually run with the pack than scoop the competition," conclude Pamela Shoemaker and Stephen Reese.[106] One reason, they say, is that exclusives do little to draw more readers or viewers. Exclusives do serve as an index of reportorial skill, but even then reporters do not want to be too far out front of the competition, Shoemaker and Reese say.

Journalists routinely check to see what stories competitors have covered, as well as the angles used in reporting the stories and the sources quoted in the stories. Scholars have noted this behavior for decades. A classic account of this behavior by journalists was written by Timothy Crouse, who traveled with the reporters who were covering the 1972 presidential campaign. Crouse found that campaign reporters relied on each other's interpretation of events—and most heavily on the lead of the news report filed by the Associated Press reporter traveling with the campaign. The reporters knew that their editors would question them sharply if their own reports deviated from the AP account, so the reporters hewed closely to whatever the AP reporter wrote. More recently, journalist Robert Parry witnessed pack journalism in coverage of the Iran-Contra scandal in the 1980s. "A principal law of pack journalism is that the more obvious a story and the less resourcefulness needed to get it, the more reporters it will draw," he said. Pack journalism arises from "the recognition that others want a story drives up its value."[107]

One example of the pack mentality among science and medical journalists occurred in 2001 when the *Lancet* published a short article that argued that there is no evidence that breast mammography leads to a reduction in deaths from breast cancer. The study at first received only limited media attention. But after the *New York Times* ran a front-page article about the study six weeks later, it became the subject of intensive news coverage and public debate.[108]

Like other journalists, science and medical journalists demonstrate a desire for certainty and consistency in their reporting. One former medical journalist described how press conferences reinforced this tendency for journalists to copy their competition: "We would go up like lambs to slaughter, and do exactly what the PR people in the institutions would want. All of these reporters, broadcast and print, would be at the press conference, and they would know that if they don't report the story today, they will be beaten by the guy sitting next to them. So everyone would rush to write the story."[109]

The journals' embargoes have this same effect, creating a type of virtual pack journalism that involves science and medical reporters across the globe, week after week. The embargo system appears to have shaped the definition of competition shared by the journalists who participate in the embargo. To them, competition means competition within the bounds of the embargo rules, such as the ability to secure comment from certain researchers or to mine the embargoed press releases to find gems overlooked by other journalists.

The embargo also capitalizes on the fact that producing a media product—such as a newspaper, television broadcast, or Web site—is difficult. It requires journalists to identify the proper kinds of news stories, collect information for them, and put them into the proper format, all under specific deadlines. Some media also operate under technical constraints that make the job even more difficult; for example, television producers often must identify video footage to help illustrate a story.

Sources of news consequently take on great importance to journalists and journalistic organizations. By definition, without sources of raw news, journalists could produce no journalistic product for publication or broadcast, just as an automobile factory could produce no cars if its sources of steel were cut off. However, news sources are not of equal value to a journalist. This is not true just of science and medical reporters. Still, science and medical journalists appear to have an atypically cordial relationship with their sources. As one science journalist put it: "Science reporters rarely display the skepticism and schadenfreude that characterizes press coverage of other subjects: politics, business, sports or even entertainment. Their relationship with their sources is almost never adversarial. Amid the cynical, spin-battered press corps, science journalists are a remarkably mild, congenial bunch."[110]

Some source bureaucracies have recognized the bureaucratic orientation of news organizations and the potential influence this gives them over the social construction of news. These bureaucracies have found ways to expand that influence, such as by providing press releases and by exerting control over journalists' access. Observing this phenomenon, Oscar Gandy argues that resource-rich institutions provide "information subsidies" to the mass media that reduce the media's cost of news gathering. An information subsidy is prepackaged information provided by a news source in a way to make it easier (or less expensive, in the sense of the use of reporters' time and effort) for a news organization to construct that information as news. When faced with an array of information sources from which to choose, journalists are likely to choose sources that are less expensive, and journalists thus would preferentially use sources whose costs are subsidized, he says. A bureaucracy that produces an information subsidy does so with a specific purpose in mind, Gandy says: "It is through the provision of information subsidies to and through the mass media that those with economic power are able to maintain their control over a capitalist society."[111]

Science journalists long have been influenced by information subsidies. Their widespread dependence on prepackaged information such as that provided through journal embargoes was noted in the first major study of science journalism, by Hillier Krieghbaum, in 1967. He wrote that "while some top-flight science reporters do go out foraging in laboratories and on campuses for news, most spend their time attending science and technical conventions, reading journals, and scanning press releases. More than in most other fields, such as politics, say, the news comes to the science writers."[112]

Although one might argue that political journalists today are much more dependent on information subsidies, science journalists' dependence on information subsidies clearly continues. In a survey of health journalists at local television stations, more than half said that their news coverage decisions were influenced by a contact from a public relations official; almost as many said that press releases influenced them. "This suggests that reporters are learning of story ideas through a 'passive news discovery process' in which reporters find story ideas without ever leaving the newsroom." Another recent critique of science and medical coverage argues that journalists often limit themselves to a press release's characterization of a study and fail themselves to plumb the study for findings that the press release's source failed to highlight. And one physician, advising journalists on how to more accurately report on medical journal articles, advised journalists to rely on one specific type of information subsidy: the editorial that sometimes accompanies a medical journal article and often explains the article's implications. "Read the accompanying editorial, if there is one," the physician advises. "It might do a lot of your work for you."[113]

One incident starkly illustrates science journalists' dependence on press releases: In late November 1996, *Science* issued its routine weekly embargoed press release describing articles in its November 29 issue. As usual, the release described some papers in great detail, including notes written by the AAAS's public relations staff that explained the papers to journalists in lay language to assist them in covering the papers. But a handful of other papers at the end of the press release were listed only by title and author, with no explanatory notes for the journalists; to use Gandy's term, the press release offered less of an information subsidy for these papers. One of those in the latter category was a paper that reported the possible discovery of ice on the moon. Journalists overlooked this paper; a check of thirty U.S. daily newspapers shows that none covered the paper when it was published. However, on December 2, the U.S. Defense Department—which had sponsored the research—called a news conference to discuss the paper. The Associated Press moved a story about the paper immediately, and newspapers followed suit with their own reporting, even though a paper from *Science* ordinarily would no longer be considered newsworthy so long after publication.[114]

Of course, American journalists are not unique in their reliance on information subsidies: One study of coverage by two British newspapers of two medical journals over a two-year period found that the newspapers covered no journal article unless it had been accompanied by a press release. Another study of the British press found that press releases produced by the *British Medical Journal* and the *Lancet* helped science journalists determine which stories to cover.[115]

Time and timeliness are very influential in the construction of news. Journalists understand news to be information that is "new" in some fashion. One way for this criterion to be met is for the informational content of the story itself to be wholly new information. Older information may still qualify as news if it can be tied to a "news peg," or an event that imparts a sense of newness to the information. Gaye Tuchman found that journalistic organizations align their news nets with source bureaucracies in part because of those bureaucracies' ability to provide news that has the desired element of timeliness. Or as the journalism historian Michael Schudson puts it: "Journalists do not seek only timely news, if by 'timely' one means 'immediate' or as close to the present as possible. Journalists also seek coincident and convenient news, as close to the *deadline* as possible." One scientist noted this propensity as early as 1919: "If you can write a scientific truth so that the principal statement of it shall be in the first sentence, and the most important words in that sentence are 'here' and 'today,' and your own name, and especially if you can write it that it would be absurd to date it at any other place, on any other date, or with any other name, then you can probably get it into the newspaper."[116]

The embargo on science journals plays on this penchant by providing a "time peg," or an impression of immediacy, for information that otherwise might not have one, allowing the science journalists to construct the news as late breaking. But such immediacy is, of course, wholly artificial. For any scientific paper being published in a scientific journal, the research—that is, the "discovery"—was made months or more earlier. Because of the journals' production deadlines, the final version of the paper had probably been accepted some weeks prior to actual publication. Under such circumstances, it flies in the face of logic to argue that a scientific paper is hot news one day and old the next. Science journalists acknowledge this, even while implicitly ignoring it in their coverage.

The choice of any particular release time will benefit certain media at the expense of others, since media publication times are scattered throughout the day. In general, embargo release times today are set so that evening television news broadcasts generally get the first opportunity to disseminate the news. This is a source of frustration among newspaper reporters, such as Mike Norris of the *Reno Gazette-Journal,* whose newspaper broke an evening embargo on a 1990 report on nuclear waste from the National Academy of Sciences. "We are sitting here at the newspaper and we feel that we are getting dumped on because we are not going to get the news out until the next day."[117]

The importance that journals place on using an aura of immediacy to promote news coverage of their articles is illustrated by the care with which journal editors choose the release time for the embargoes. The timing of the embargo's expiration, generally in mid-to late afternoon on the U.S. East Coast, is designed to maximize the prospects for coverage by evening television news broadcasts by making the research seem to be breaking news; if the embargo time instead were set to give morning newspapers the first chance to report the research, television broadcasts later that day would be unlikely to cover the research because producers would not deem the research sufficiently new and immediate. Although network broadcasts cover science news very rarely, scholarly journals wish to do their best to get any coverage that they can.

Of course, journalists use criteria other than timeliness to determine what stories to cover. When discussing news judgment, journalists and journalism teachers often explicitly or implicitly invoke the notion of "news values," or a set of standards of newsworthiness against which potential news can be judged. Media theory suggests that news values shape how journalists devote their attention to events around them. Journalists and their organizations "look" only for stories that are deemed newsworthy a priori according to agreed-upon news values.[118]

There is considerable agreement among mass communication researchers regarding the news values that are in common use by journalists. In their pioneering work, J. Galtung and M. Ruge identify eight major factors that influence the selection of foreign news. These include the extent to which a potential news story conforms to the production schedule of the media organization, the size of a news event, the degree of ambiguity in interpretation of the event, the "meaningfulness" or cultural proximity of the news, the degree of consonance between the news and previous views about a nation or cultural group, the unexpectedness of the news, the degree to which the story has been covered in the past, and the availability of competing news stories. Synthesizing research on news values, Pamela Shoemaker and Stephen Reese have proposed six essential news values: prominence or importance, human interest, conflict or controversy, the unusual, timeliness, and proximity.[119]

Science journalists themselves enunciate criteria for assessing newsworthiness of potential science news stories that are similar to these. For example, Boyce Rensberger, the former science reporter and editor for the *Washington Post,* lists five criteria for science news that, he says, newspaper science writers use:

1. "Fascination value," or the imparting of interesting information. Rensberger states that dinosaurs, black holes, human evolution, and animal behavior are topics that are high in fascination value.
2. The size of the "natural audience" for the topic. Cancer and the common cold rank highly on this criterion, he says, because many people either have or fear contracting these illnesses.
3. Importance, or "whether the event, or finding, or wider knowledge of the event or finding is going to make much of a difference in the real world,

especially in that of the average newspaper reader," he writes. "AIDS is important, bunions are not."

4. Reliability of the results. The "single most useful guideline" to assessing reliability, he writes, is whether a research finding has been peer-reviewed.

5. Timeliness. "The newer the news, the newsier it is," writes Rensberger.[120]

There is ample evidence that science and medical journalists take such factors into account when choosing what journal articles to cover. In one analysis of newspaper coverage of the *Journal of the American Medical Association* and the *New England Journal of Medicine,* four researchers rated the importance of all articles published in the journals in 1988 and then examined coverage of the articles by ten major U.S. newspapers. Journal articles that were more important also received more, and more prominent, coverage than less important journal articles. In another analysis of news coverage of both journals, a researcher found that journal articles with implications for readers' lifestyles (such as the relationship between walking and health) received more coverage than journal articles that were focused on medical issues out of the control of readers (such as the results of a test of an experimental drug). Also, studies that cited a specific demographic group for which the study was relevant—for example, smokers—received more newspaper coverage than journal articles that made no specific connection to a demographic group.[121]

Other evidence that journalists take importance into account in their medical coverage comes from a study of newspaper coverage of cancer, heart disease, AIDS, and Alzheimer's disease from 1977 to 1997. Two researchers found that the volume of coverage reflected the rate at which Americans died from each of the diseases. "When mortality takes sharp downturns or upturns, news coverage does too."[122]

Another key element of newsworthiness is whether the press release suggests the existence of a local angle for a story. Journalists often ignore news releases that do not suggest a local angle. Under the conventions of journalism, a story is most newsworthy when it has a local angle. For this reason, journalists often ignore news releases that do not suggest a local angle. Journalists' assessments of newsworthiness may also be influenced by the way in which the journals' press releases present information to them. For example, there is evidence that science journalists, trained in the inverted-pyramid style of newswriting (in which the most important information is at the beginning of the story, followed by increasingly less important information), assume that the journal articles that are listed first in a press release are more important: in one recent study of news coverage of journals, journal articles that appeared early in a journal's press release were covered more frequently by newspapers than journal articles that appeared toward the end of the press release.[123]

A second mechanism that may be influencing journalists' construction of news about research described in a press release is the length of the press release's description of a journal article. In journalistic conventions, a more important story receives more space. By the same token, a journal citation with a long description would be deemed more important than a journal citation with a short description. Support for the plausibility of this hypothesis in one content analysis of news coverage of a university's press releases: releases that were more than three pages long were more likely to be covered than those that were three pages or less in length.[124]

The preceding discussion suggests a simple model to predict a particular newspaper's decision making regarding news coverage of a specific journal article. That model—along with data from an analysis of journal coverage by twenty-five U.S. daily newspapers from June 1997 to May 1998—is depicted in figure 1.1.[125] On the left-hand side of the figure are two qualities related to how the article is described in the journal's press release. The top one is "Depicted Newsworthiness in PR," which represents how generally newsworthy the press release describes the journal article as being, through

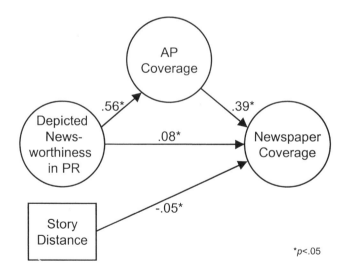

Figure 1.1 Results of a Content Analysis of Coverage of Four Journals by Twenty-five Daily Newspapers. Note: The numbers designate the estimated standardized coefficients among elements of the model.

the length of the write-up, the write-up's position in the press release, and the number of news values in the press release. Below that is "Story Distance," which represents how close the researchers are geographically to the newspaper in question. In the center of the figure is "AP Coverage," which represents the quantity and nature of the Associated Press's coverage of the journal article, as measured by the number of stories, the average length of those stories, and the number of news values in the stories. And on the right-hand side of the figure is "Newspaper Coverage," which represents the quantity and nature of a newspaper's coverage of a specific journal article, as measured by how long the newspaper's article was (if it ran an article at all), whether it appeared on page 1 or inside the paper, whether it was written by a staff writer or came from a wire service, how many photos or graphics accompanied the story, and how many "news values" the newspaper article described the journal article as having.

In the figure, a value of 1.00 or minus 1.00 would mean a perfect causal relationship (positive or negative, respectively), and a value of 0 would mean no causal relationship at all. In the analysis of newspaper coverage, the principal direct determinant of newspaper coverage of a particular journal article was coverage of the journal article by the Associated Press. This does not mean that newspapers in the study always ran the AP's story about a given study; newspapers often ran staff-written stories or stories provided by other news services. But coverage by the Associated Press appears to be an important factor in determining whether a given study was covered by a newspaper and, if so, the fashion in which it was covered. If the AP is covering a particular study, the AP article is likely to be carried by many other media outlets, so any given media outlet is itself likely to use the AP story or one from a staff journalist or other wire source to avoid missing a story that will be commonplace in other newspapers. (One possible objection to this conclusion is that the newspapers and the Associated Press independently use similar standards in deciding about news coverage. Although the statistical argument is complex, this alternative does not fit the data from the content analysis as well as the figure does, meaning that it is statistically less plausible.)[126]

The figure also shows that other influences on newspaper coverage had only a very small effect. For example, story distance had little direct impact on newspaper coverage. One possible reason is that science and medical reporters and editors may view proximity as largely irrelevant to the newsworthiness of scientific findings, because those findings (such as the effectiveness of a medical treatment) are relevant locally even if the study was conducted at a great distance.

Also, the press release's depiction of the newsworthiness of a journal article had little direct influence over newspaper coverage of that journal article. But it would be an error to jump to the conclusion that journal publishers are wasting their time and effort in supplying information subsidies to journalists, because the Associated Press's coverage of journal articles was strongly influenced by the press releases, and the AP in turn influenced newspaper coverage. Although at first blush it might seem that the journals' public relations efforts were largely ineffective, in fact they were effective indirectly, through their influence on the wire service.

Conclusions

In the United States, most science and medical journalism is performed by specialist journalists at large daily newspapers and the Associated Press, although television journalists do cover medical issues to a degree, and Web sites are a new entrant to science and medical journalism. Science and medical journalists tend to focus their reporting energies on two segments of the scientific-communication process: scientific conferences and a handful of elite journals. These journals regulate press coverage through rules restricting scientists' communication with the press and an embargo arrangement with journalists. The general public's interest in the news about science is modest; interest in medical news is somewhat higher. Public opinion research suggests that individuals who are more knowledgeable about science also have more favorable attitudes toward science, a conjunction that has encouraged the scientific community to promote more news coverage about scientific research.

2

A Brief History of Embargoes
in Science Journalism

A century ago, there was no embargo system for disseminating news about scientific research to science journalists. One reason is that there were no science journalists, at least in the sense in which that term is used today; another is that the publishers of scholarly journals likely saw little value in promoting media coverage of their publications. The story of the development and spread of embargoes on scientific and medical research provides a unique window on the growth of both institutional science and journalism in a century during which both mightily strove to demonstrate their importance to society.

Reporting on science and medicine has been part of American journalism from the start: the first American newspaper, *Publick Occurrences,* published in 1690, carried two paragraphs on smallpox, "which might be cited as the first American newspaper reporting of science news." Six decades later, Benjamin Franklin published an account of his landmark experiments with lightning in his *Pennsylvania Gazette.* But overall, prior to the nineteenth century, newspaper reporting of science was spotty and unsystematic. The expansion of the newspaper industry in the early 1800s, through the development of the so-called penny press, stoked editors' interest in news about science and medicine. Stories in this era included agriculture, medicine, the exploration of Africa, and inventions such as the telegraph, telephone, and electric lights. Public interest in science and technology—and consequently, journalists' desire to cover such topics—grew sharply after the start of the twentieth century, and particularly so after World War I, which was heavily influenced by technological innovations such as aviation, submarines, chemical weapons, and radio communications. In addition, the *New York Times* thrust science into the attention of the American public in 1919 when the paper reported that astronomers who had measured starlight during a solar eclipse had concluded that the measurements proved Albert Einstein's theory of relativity. Few had even heard of Einstein before the *Times*'s report.[1] Still, the coverage of science generally steered clear of reporting of the publication of scientific papers and of the presentation of papers at scholarly meetings: an examination of newspaper coverage of a geology conference in 1913 found that the newspapers collectively published only 3,160 words on papers presented at the conference, compared to 14,420 words on nonscientific aspects of the proceedings, such as addresses by government officials, and 10,310 words on the conference's social functions.[2]

Birth of an Idea: The 1920s

The 1920s marked an important turning point for science journalism, in the creation of the craft itself and the public relations apparatus that would evolve into the full-blown embargo system as it is known today. In particular, the first newspaper science editors were named in this period, including Alva Johnson for the *New York Times,* David Dietz for Scripps-Howard newspapers, and John J. O'Neill for the *Brooklyn Daily Eagle.*[3]

Another influential force for embargoes in early science journalism was Science Service, a not-for-profit news agency founded in 1921 with financial support from E. W. Scripps in an effort to promote wider popular understanding of science. Science Service distributed news stories to subscribing newspapers by mail and telegraph, and the founding editor, Edwin E. Slosson, made clear from the very beginning that Science Service planned to arrange for extensive embargoed access to scientific reports. In his discussions with scientists, Slosson repeatedly linked this access to the goal of wider appreciation of science, which he argued both was a social good in itself and would promote wider support for science in U.S. society. For example, in his announcement of the formation of Science Service, Slosson wrote:

> In a democracy like ours it is particularly important that the people as a whole should so far as possible understand the aims and achievements of modern science, not only because of the value of such knowledge to themselves but because research directly or indirectly depends upon popular appreciation of its methods. In fact the success of democratic government as well as the prosperity of the individual may be said to depend upon the ability of the

people to distinguish between real science and fake, between the genuine expert and the pretender. . . . The editor of Science Service desires to receive advance information of important researches approaching the point of publicity in order to arrange for their proper presentation in the press.[4]

Slosson made this point as well in letters to leaders of scientific organizations, in which he sought advance access to scientific papers and publications, linking that to the institution's own self-interest. For example, in one February 1921 letter to John C. Merriam, president of the Carnegie Institution of Washington, Slosson requested that Science Service be supplied with extensive advance information about the institution's activities, such as proof sheets of forthcoming publications. In a subsequent letter to the institution, Slosson made clear the scientists' interest in supporting embargoes: "The only way to prevent the misinterpretation of the announcements of a scientific discovery is to have prepared in advance for simultaneous release a popularly written explanation of its meaning and significance."[5]

Science Service in turn provided its subscribing newspapers with news reports under embargo. Its first such embargoed story reported details of a paper scheduled for presentation to the National Academy of Sciences, so the story was embargoed for release in morning editions of April 27, 1921.[6]

One concrete step that Science Service took to promote embargoes was to request advance copies of papers from all presenters at the annual conference of the American Association for the Advancement of Science in December 1921 in Toronto. One copy went to Science Service so that its reporters could produce embargoed stories on the conference; the other copy was given to the local publicity chairman. The value for journalists of having embargoed access to these scientific papers was underscored a year later, in 1922, when *New York Times* reporter Alva Johnson used such access to produce news coverage of the conference that won a Pulitzer Prize in 1923.[7] The embargo system developed by Science Service continues in a modified form today, at meetings both of the AAAS and of many other scientific societies.

Other scientific societies also began using embargoes to reach the new science journalists. In 1923, James T. Grady, public information officer of the American Chemical Society, obtained advance copies of scientific papers to be presented at the chemical society meeting and distributed news releases about the papers in advance of the meeting under an embargo.[8] The American Medical Association, under the leadership of Morris Fishbein, the flamboyant and politically powerful editor of the *Journal of the American Medical Association* from 1924 to 1950, also embraced the use of embargoed materials for its conferences. In a 1936 speech, Fishbein described the system followed by the medical association: The association would request all speakers to submit abstracts of their presentations at least three weeks before the conference. The AMA's staff vetted the abstracts, looking for papers that both would be of public interest and would meet a test of "established evidence"; for these, the AMA prepared press releases with embargo times keyed to the time of presentation, which were distributed to the media. During the conference itself, Fishbein said, meetings were held twice every day with journalists

> so that questions may be asked and answered and any material developed by the reporters may be suitably appraised before released to the public. We find the press invariably anxious to cooperate and ready to limit their releases to material approved by the publicity organization. Reporters are instructed not to accept interviews from any contributor to the program until such interviews have been suitably approved by the publicity organization. In this they also cooperate.[9]

Fishbein himself played a major role in this process. One history of the National Association of Science Writers recalls: "Many an NASW member will recall those wondrous 'tours' before each convention where Dr. Fishbein briefed the reporters and took them on the tour of the scientific exhibits and in his staccato, breezy style gave them advance knowledge of the importance of the discoveries on display."[10]

The origins of journal embargoes are less clear, but they have their roots in this period as well. Officials at the American Medical Association believe that Fishbein originated the notion of providing journalists with embargoed access to journal articles. Fishbein writes in his autobiography that he allowed Howard W. Blakeslee, a Chicago-based reporter for the Associated Press and later its first science editor, to visit the AMA's Chicago headquarters each week to read page proofs of the impending issue of the journal. Unfortunately, Fishbein is hazy on key details, such as when the arrangement began, who proposed it, how and when it was expanded to other reporters, and even on the details of the embargo itself. In his autobiography, Fishbein recalls, "I developed a relationship with Howard Blakeslee, who represented the Associated Press in Chicago. He came to the office each Wednesday to receive advance proofs of important material which was then circulated around the world." However, at another point in the autobiography, Fishbein writes: "When Howard Blakeslee was assigned to Chicago by the Associated Press, he had arranged to come each Thursday to AMA headquarters where I conferred with him and pointed out all new scientific developments."[11]

These two descriptions differ in more than just whether Blakeslee's weekly consultation was on Wednesday or Thursday. The first suggests that the system was developed at Fishbein's instigation; the second suggests that Blakeslee was the driving force behind the embargo arrangement. Both descriptions suggest that Fishbein attempted to place his own interpretation on Blakeslee's coverage of the journal's contents. Certainly, Blakeslee was skilled at insinuating himself into scientific circles, as Kent Cooper, general manager of the AP from 1925 to 1948, notes about Blakeslee in his own autobiography: "Besides being a good reporter, he was friendly and diplomatic with the scientific and medical men upon whom he knew he had to rely. He got acquainted with them first as friends, and on the basis of friendship he got them to talk."[12]

Fishbein's efforts with Blakeslee were noted by other journalists. Arthur J. Snider, science writer for the *Chicago Daily News,* later cited the embargo arrangement in a letter that Snider wrote to support honorary membership for Fishbein in the National Association of Science Writers: "It is quite likely that Howard Blakeslee's career as a science writer was spawned by Fishbein's tutoring."[13]

James Stacey, formerly the director of media and information services for the AMA's Washington office, says that the embargo served both men's needs. "A growing public interest in medical information was being demonstrated, and Fishbein wanted to make the information accessible and understandable. That goal was shared by Blakeslee, who was routinely able to question Fishbein and the published researchers about the meaning and implications of their work." However, Stacey ignores the possibility that the journal and the AMA benefited from the added publicity that the embargo arrangement presumably attracted for the journal's articles and Fishbein.[14]

Although Fishbein saw that the embargo did benefit his journal, he was unwilling to share those benefits with the institutions at which scientific research was conducted. For example, the director of the Massachusetts General Hospital wrote to the AMA in 1937 to seek advice on a request from the Associated Press that the hospital provide the wire service with advance access to research papers "on the distinct understanding that it be held for release until it is presented to the profession." Replying on behalf of the medical association, Fishbein strongly hinted that researchers could jeopardize their chances of publishing research shared with the wire service: "As editor of the *Journal of the American Medical Association* I would be inclined to resent release of such publicity in any way except through The Journal, and I believe editors of other publications would be similarly resentful."[15]

Lobbying by Journalists: The 1930s and 1940s

For the embargo to work, scientists had to provide texts of their papers in advance, but initially few did. For example, at the December 1933 annual convention of the AAAS, only about one-fifth of the scientists submitted papers in advance for press use.[16] Thus, after twelve leading science journalists established the National Association of Science Writers in 1934, the organization's leaders set themselves the task of hectoring scientists into providing those advance copies.

For example, David Dietz, science editor of Scripps-Howard Newspapers and a former president of the NASW, told the 1936 AAAS conference that advance copies were essential for high-quality news coverage of the meeting:

> Each day, there are some twenty or thirty sectional meetings in session. The important papers on any day may be read in various sections meeting in widely separated buildings. Even if all the papers in which a particular newspaperman was interested were read in one meeting, the time element would still have to be taken into consideration. A reporter cannot sit through a long session, return to his typewriter and still get his account into the day's newspapers. He must have the papers in advance.[17]

The journalists' interest in getting advance copies was strengthened after five science reporters—Dietz, Blakeslee, Gobind Behari Lal of Universal Service, William L. Laurence of the *New York Times,* and John J. O'Neill of the *New York Herald Tribune*—shared a Pulitzer Prize in 1937 for their coverage of a 1936 science conference marking the tercentenary of Harvard University. Harvard officials had aided the reporters immensely, including providing advance copies of papers and arranging press conferences with speakers. Laurence later told an oral historian that the embargoed materials were the key to his coverage. "When I got there I got a package of about 70 papers—a regular encyclopedia—given to me, together with very fine interpretations, [or] summaries, because some of them were very technical papers written by the Harvard faculty. And that helped me no end. Otherwise I couldn't have covered it."[18] Laurence spent each day of the conference writing news stories for the *Times* based on the advance texts as well as information from press conferences and interviews. It is unclear from his comments whether Laurence spent any time at all in actual conference sessions.

The 1937 Pulitzer Prize was not lost on leaders of the scientific community. Sidney S. Negus, who later would run a pressroom at AAAS conferences, cited the Pulitzer in a 1940 memorandum urging officials of the American Pharmaceutical Association to provide journalists with advance copies of papers for its 1940 meeting in Richmond, Virginia. "It is best to put the facts in the hands of *trained* science writers one can depend upon and give them *time* enough to put their material in condition for the papers they represent," Negus wrote. "Otherwise, we will either see no national newspaper reports on the meeting or else what we do see will be badly reported—what a mess an untrained reporter can make of a scientific report!!"[19]

Austin Clark, head of the Smithsonian Institution—and an observer at the founding meeting of the NASW—organized a pressroom for conferences of the American Association for the Advancement of Science in which advance texts were distributed to journalists. Clark believed that it was in science's own institutional self-interest to help journalists publicize scientific findings, as he argued in an article titled "Science Progress through Publicity" in one of the AAAS's journals, *Scientific Monthly.* Clark warned that "unless any given group within a social unit is recognized as contributing to the material or spiritual welfare of that unit, sooner or later it will be in danger of elimination. ... Our duty to the community in which we live, to science, and to ourselves is to take the public completely into our confidence and to provide the interpreters—the science writers—with all the material they can use."[20]

The journalists even extended advice on how scientists should write abstracts of their papers in order to better catch the eyes of journalists. Herbert Nichols, science editor of the *Christian Science Monitor* and president of the NASW, wrote to scientists in 1947 in the journal *Scientific Monthly* that science journalists were assigning letter grades to the abstracts—and he warned that those with poor grades got little news coverage. "In the A.A.A.S., it is frequently difficult for the press to find enough in the abstracts to meet its need. Too many speakers send in no abstract at all, furnish only the title, or write a single sentence intended either to discourage the reporter or cause him to seek out the author."[21]

Progress in converting scientists was slow. In 1949, Barrows Colton, president of the NASW and a staff writer at *National Geographic,* wrote a lengthy article in the journal *Scientific Monthly* to explain in detail how deadlines work for morning and evening newspapers—and hinting that scientists might regret it if they didn't provide advance copies of their papers.

> It is of vital importance that speakers at scientific meetings provide complete copies of their papers in advance to the press, if possible at least 24 hours ahead of the time of delivery. Then the reporter can write his story at some

leisure, with all the facts before him, and he has time to seek out the speaker for additional particulars if necessary. If he has no copy of the paper, he must try to waylay the speaker whenever and wherever he can to learn what he is going to say, bothering him perhaps at inconvenient times, and making hasty notes that are far less satisfactory than a copy of the paper, and more likely to result in an inaccurate or incomplete story.[22]

The AAAS also used the specter of inaccurate reporting to goad scholars into cooperating with journalists. For example, the press officer for the association's 1949 conference, in New York City, wrote the conference's officials four months in advance to request 100 copies of each paper and 120 copies of the abstract of each paper for distribution to journalists. "We could let the science writers figure it all out for themselves, attend the various meetings and rush through spot stories but, under this plan, the Association and Societies would surely get inadequate coverage of the meetings and many inaccurate reports would be published all over the world which would be tremendously embarrassing to many presenting papers."[23]

Despite journalists' incessant emphasis on embargoed access to conference papers, it is clear that they continued to rely heavily on scientific journals as well for their news stories. An analysis of the science sections of the Sunday editions of the *New York Times* and the *New York Herald Tribune* in 1946 and 1947 found that 45 percent of their stories came from journal articles, making journals used as frequently as any other source. *Science* was the most frequently quoted published source of science news in those two papers. The *Journal of the American Medical Association* was second, and *Nature* was third. The *New England Journal of Medicine* was a distant also-ran, one of nineteen publications that were quoted less than three times during the analysis period.[24]

Scientific Associations Respond: The 1950s

The hounding from journalists did the trick. Scientific associations began assembling advance information for journalists who were planning to cover their conferences, and scientists increasingly cooperated by submitting papers and abstracts. In 1956, for example, the American Medical Association began sending journalists advance abstracts of papers to be presented at its annual meeting, and the conference's pressroom was equipped with a photocopier to duplicate the full texts of papers. By the following year's conference, AMA staff had devised a numbered index for conference papers so journalists could locate them more easily. By this time, the AMA also was sending reporters weekly press releases describing research published in its journals.[25]

Similarly, the American Association for the Advancement of Science had begun distributing extensive advance information about its annual conference to journalists. For its 1957 meeting in Indianapolis, the association notified journalists from the home cities of all speakers at the conference as well as the speakers' universities, companies, or agencies. A pressroom distributed texts and abstracts of papers, and the head of public relations for the conference estimated that print journalists alone filed one hundred thousand words of news coverage during the weeklong event.[26]

Other associations followed a similar practice. In late 1950, the American Association for Cancer Research began providing journalists with page proofs of its journal, *Cancer Research,* giving journalists about two weeks to prepare stories before the journal articles were formally published. The Archaeological Institute of America similarly distributed page proofs of its journal, *Archaeology.* After Science Service failed to report on any of that journal's articles for some time, the editor wrote to Watson Davis to complain. "Some of our articles have been too good to miss," wrote the editor, Jotham Johnson. In his letter, he went on to describe four noteworthy articles in the accompanying page proofs.[27]

The scientific associations' embrace of embargoes may be explained, at least in part, by the fact that in the 1950s, the scientific establishment started to focus on what one scholar has called "the 'problem' of popular science"—an understanding among scientific leaders that the popular media should serve as a conduit for information transmitted by scientists to the general public. "The goal is to minimize media interference so as to transmit as much information as possible with maximum fidelity."[28]

In this period, embargoes were fast developing into the rule rather than the exception, as was well illustrated by science journalists' anger over a 1955 incident in which the University of Michigan released a long-awaited study on the efficacy of the Salk polio vaccine without providing embargoed copies in advance. Arthur Snider, science writer for the *Chicago Daily News,* recalled that "they loosed cries of anguish on learning that they would be given the highly secret Francis report at the same time as the hundreds of general assignment reporters. As science writers, they felt that they were entitled to the privilege of advance perusal before being obliged to transmit their stories in a matter of minutes—a strange change of attitude from their police reporting days when they covered stories under much greater handicaps."[29]

Reporters received copies of the report about an hour before a presentation in Ann Arbor by the scientist who headed the study. The pressroom at the university was a study in bedlam, with the university's public affairs officer climbing atop a table to hand out copies of the report. One science journalist later said the journalists' rush for copies of the study "was like D-Day at Normandy." Some journalists complained that the National Broadcasting Company broke the hourlong embargo by airing the study's conclusion on the *Today Show*. The university's public affairs officer later mused that "if he had it to do over again . . . he would not have given out the reports to the pressroom first, because few newspeople actually attended the conference."[30]

Journals remained an important source of news. Arthur Snider of the *Chicago Daily News* estimated that 10 percent of science news stories published by his paper came from scientific journals. Journal publishers recognized this phenomenon. Dael Wolfle, a senior official at the American Association for the Advancement of Science, wrote in one internal report that "*Science* is not intended for a large audience, but some of the information published in it merits wide dissemination. By sending advance copies to a number of science writers (and galley proofs to a few), we assist in the preparation of a considerable number of newspaper and magazine accounts of recent scientific happenings."[31]

With the wider availability of embargoed information—particularly at scientific conferences—journalists began to find themselves in conflict with one another over who would benefit from the terms of the embargo. Initially, these conflicts pitted evening newspapers against morning papers. For example, a 1953 conference of the American Psychiatric Association in San Francisco was disrupted when the editor of an evening paper ordered his reporter to file stories on presentations the evening before they were to be made. Other journalists at the conference banded together to deny the paper any scoops. "It got back into line."[32]

Throughout the 1950s, members of the National Association of Science Writers complained about problems with embargo release times and debated what to do. The organization surveyed its members in 1954 on the issue; tellingly, survey questions were limited to the time at which embargoes on conference news should be lifted and did not seek opinions on whether advance news should be provided at all. Based on the survey, the NASW recommended that conference papers delivered before 1 P.M. should be released for coverage in afternoon newspapers that same day; presentations later in the day should be held for the next day's morning editions. The NASW also declared that "extraordinary news" should be publishable as soon as announced but studiously avoided defining the term, explaining that "social pressures would restrain many reporters from using the 'extraordinary' news plea too frequently."[33]

However, this did not settle the problem. Disputes among journalists again erupted at a 1959 seminar for science writers sponsored by the American Cancer Society, when the *Chicago Tribune* decided to violate embargo times on papers presented to the writers by researchers. In a statement published in the paper, *Tribune* editor Don Maxwell accused other journalists of agreeing to "hold back reports of important medical discussions as long as 24 hours." The cancer society expelled Roy Gibbons, the *Tribune*'s science editor, from the conference.[34]

New Conflicts: The 1960s

A new theme in disputes over the release of scientific findings started about 1960, when Samuel Goudsmit, editor of *Physical Review Letters,* a physics journal, began to fret that the mass media were reporting scientific reports before they had been vetted through peer review. He also worried that such media reports also undercut the rationale for the journal's own existence. Goudsmit declared to his readers that he would refuse to publish any paper that had already been publicized in the mass media: "Scientific discoveries are not the proper subject for newspaper scoops. . . . Formerly crackpots often made the front page with their spectacular stories, and this still happens occasionally. We are sure that our authors do not wish to be confused with these pseudo-scientists in the minds of the public. This can be avoided by using the right publicity channels which will give these stories an authoritative stamp of reliability and the proper dignity."[35]

Goudsmit would later claim that science journalists had fully supported his policy, and, indeed, several science journalists interviewed by a scholar in 1964 voiced support for the policy. For example, Ann Ewing of Science Service said that journalists could risk publishing invalid scientific findings by printing stories before peer review by a journal was complete. "The time cut is not worth the possibility of inaccuracy," she said.[36]

Nonetheless, Goudsmit's stand immediately caused problems for science journalists seeking to cover scientific discoveries made by space probes launched by the National Aeronautics and Space Administration. Tensions already were high between science journalists and a variety of government agencies that were pursuing high-profile scientific endeavors at public expense, such as NASA, the National Science Foundation, and the National Institutes of Health. NASA, for example, was legally mandated to disseminate information on its discoveries, but journalists complained that the agency instead controlled its scientists so tightly that most information was stifled.[37]

Indeed, secrecy was rampant in NASA's early years. In contrast to NASA's current practice, the space agency classified the planned dates of upcoming human space flights and provided little advance information on the scientific experiments to be conducted during them. The first head of NASA, T. Keith Glennan, wrote in his memoirs that he was concerned "that the papers would make a circus of our Project Mercury as we approach the time when a man will enter the capsule to undertake a flight." In December 1960, Glennan discussed his concerns with Benjamin McKelway, editor of the *Washington Evening Star* and president of the Associated Press, and Russell Wiggins, editor of the *Washington Post.* Both journalists argued against NASA's view that prelaunch publicity would create a sense of letdown if a launch were postponed or would be seen as an attempt by NASA to lay the political groundwork for failures in some of the *Mercury* experiments. The editors suggested that NASA distribute a package of background information on the experiments, including a frank discussion of their potential problems, in advance of the mission. This information should not be embargoed until launch time but rather should be approved for immediate use by the journalists, the two editors argued. Glennan took their advice: NASA agreed to distribute advance information about upcoming flights—without indicating the planned launch time. "It will be embargoed for only three days, to permit it to reach an enlarged number of potential users, and then will be for use at the option of the receiving newsmen. The information also will clearly and specifically point up the research and development nature of the experiment. In this way we hope to accomplish desired improvement in communication with the public about the experimental nature of our undertakings."[38]

In the midst of such struggles between journalists and space-agency bureaucrats, journal editors' insistence that they would not publish scientific findings that had been disseminated in the public press gave NASA scientists an additional rationale to postpone public release of scientific data and findings. However, Robert C. Toth of the *Los Angeles Times* raised a new argument in favor of immediate release: the research was publicly funded and therefore should be open to public scrutiny. At his insistence, NASA officials agreed to a compromise. Major findings, such as the first photograph of the far side of the moon, would be released to the news media immediately, even before submission to a journal. Lesser findings would be submitted first to the journal *Science,* and they would be released to the press as soon as the journal accepted them for publication, without waiting for the actual scientific publication to occur. "I guess we've gotten something of a victory, if it holds up," Toth concluded. This conflict illustrates the power of embargoes to aim journalists' attention in certain directions—and away from other directions; NASA's embrace of the embargo system can be seen as part of a larger public relations strategy to direct science journalists' attention toward NASA's scientific accomplishments and away from management and political issues at the space agency. Bruce Lewenstein argues that this strategy left journalists unprepared to cover management problems at NASA when the explosion of the shuttle *Challenger* would make them painfully evident two decades later.[39]

Other U.S. government agencies, eager to publicize their research, also found that embargoes were a useful tool in drawing publicity. For example, in June 1964, the National Science Foundation distributed an embargoed press release reporting findings on astronomical observations made from a telescope carried into the stratosphere by balloon. The embargo was timed to produce stories in the Sunday newspapers. "The release was sent to science writers far enough in advance to give them time to secure interviews, comments and background on the importance of the telescope's findings." An increasing number of scientists, too, were beginning to perceive that they could benefit from news coverage of their research. "With a realization that more money for research, improved status, and a better public image are among the benefits that can flow from better understanding, he has become more receptive to active participation in public relations activity," concluded one science news critique written in 1962. Journals, too, saw benefits from embargoes: from October 1965 through January 1966, *Science* was cited at least 295 times in U.S. newspapers and magazines.[40]

But some scientists remained reluctant to cooperate. Victor Cohn, the president of the National Association of Science Writers and a science reporter for the *Minneapolis Tribune*, remonstrated: "We depend on your papers and summaries. Far too few of you send them in. Most of all, we want to be able to read your complete paper. This gives us the best chance to report clearly and accurately. . . . We know, of course, that sometimes you finish writing your paper only at the last minute. Even so, we plead, bring a copy to press headquarters."[41]

By the 1960s, science journalists shared a widespread—and apparently uncritical—acceptance of journal and conference embargoes. For example, a 1963 textbook on science writing advised budding science journalists: "As your knowledge of science increases and your circle of contacts grows, you learn of more and more developments that you want to report right now.

But you will run up against an old scientific law—the scientist must report to his peers, either at a scientific meeting or through a scientific journal, before reporting to the public through the mass media. The only thing you can do now is abide by the law, no matter how much you may dislike it."[42]

And by 1963, a scholar observed the important role that embargoes played at medical conferences: "Press headquarters form an integral part of every medical meeting today with advance press releases and background information provided every attending medical writer and editor to assure accurate coverage of medical news, information and education." The embargo, the scholar wrote, is "structured to protect the informal code of the medical profession that medical information is not official or available for comment in the mass media until publication is made of a medical paper in a medical journal or at a medical convention."[43]

Use of embargoed material became so widespread that by 1965, a journalist could write—apparently satirically—that, at scientific conferences, "one of the biggest headaches in pressrooms is the 'textual deviate.' This is the scientist who makes an advance text of his talk available days ahead of time, then gets up at the meeting and talks off the cuff."[44]

The widespread dependence by science writers on prepackaged information such as that provided through journal embargoes was noted in the first major study of science journalism, by Hillier Krieghbaum in 1967. He wrote that "while some top-flight science reporters do go out foraging in laboratories and on campuses for news, most spend their time attending science and technical conventions, reading journals, and scanning press releases. More than in most other fields, such as politics, say, the news comes to the science writers." Nate Haseltine, the editor of the newsletter of the National Association of Science Writers, acknowledged as much that same year when he argued that science journalism had evolved into a "no-scoop climate." Haseltine said: "Under today's working conditions it is virtually impossible to score real scoops on scientific developments. After all, we all get the same journals and releases from the same sources."[45]

In this era, the National Association of Science Writers prepared a guidebook for press officers that strongly endorsed the distribution of embargoed copies of conference papers, but the way in which the argument was framed indicates that broadcast coverage was not a major factor. "Most newspaper writers must prepare their stories well in advance of their newspaper deadlines. This is especially true of science stories, which often take a long time to prepare. This is the primary reason why advance texts are so important. For example: If afternoon newspaper reporters must wait until a speaker presents a paper at 11 a.m. before starting to write and transmit their stories, most of them will miss their early edition times."[46]

But as evening newspapers began to die off and newspaper science journalists began to see television journalists as their principal professional competitors, arguments over embargo release times turned to whether television or newspapers should be favored by the release times. The 1967 conference of the American Association for the Advancement of Science in New York was particularly rancorous on this score. "Television repeatedly broke these times in New York while many writers tried to observe them. We faced a mild chaos," wrote Robert Cowen, NASW president and science editor of the *Christian Science Monitor.* Three months later, representatives of the NASW and AAAS met to try to prevent further problems. The journalists' proposals included an expansion of the embargo system, under which papers would be distributed two weeks in advance of the meeting and "textual deviates" would notify journalists in the pressroom that they had departed from their texts. The issue of television journalists violating the embargo times apparently was not discussed.[47]

At this time, science journalists were beginning to complain bitterly about the tactics used by scientists and commercial public relations firms to manipulate news coverage, including embargoes. But these complaints never cited embargoes on journals or conference papers, which suggests that journalists did not recognize—or chose not to recognize—the public relations value that these embargoes served for journal publishers and conference organizers.

For example, in 1963, one of the architects of the embargo system—Watson Davis of Science Service, who had argued so forcefully at the 1937 AAAS meeting for advance access to scientific papers—objected that companies and advocacy groups such as the American Cancer Society were using public relations tactics such as embargoes in a "Madison Avenue approach" to selling science. Issuing material under an embargo, he complained, "even though the information is not too new and pertinent, forces the science writer or editor to give more attention that he otherwise would to an announcement for fear that other newspapers will issue a report and he will be left in the lurch."[48] But Davis did not cite journal or conference embargoes in his complaint.

And it appears that many science journalists did not recognize that they themselves manipulated news flow to their advantage through embargoes. For example, in 1966, embargoes were placed on news coverage of an annual briefing held for science writers by the Council for the Advancement of Science Writing, an organization closely linked to the National Association of

Science Writers. The embargo, under which presentations from the conference were held for forty-eight hours, was imposed to address complaints from reporters at the previous year's briefing who had to miss conference sessions in order to file stories on deadline.[49]

In one respect, the 1960s drew to an end much as they began, with a journal editor asserting control over interactions between scientists and journalists regarding research that had not yet been peer-reviewed. In 1969, Franz Ingelfinger, editor of the *New England Journal of Medicine*, became upset when a scientist submitted a paper to his journal that had already been described in detail by the *Medical World Tribune*, a medical trade newspaper, covering a medical conference. The journal had already stated that it would publish only papers that had not been published or submitted for publication elsewhere. But the incident spurred Ingelfinger to issue a more specific statement that he would publish only papers that had been "neither published nor submitted elsewhere (including news media and controlled-circulation publications)." Ingelfinger exempted news media coverage of talks by medical researchers at scientific meetings but clearly signaled researchers that they should use extreme caution in conducting interviews with reporters that would expand on the often sketchy information contained in such talks: Ingelfinger declared that he would refuse to publish a paper "if the speaker makes illustrations available to the interviewer, or if the published interview covers practically all the principal points contained in a subsequently submitted manuscript."[50]

Unlike Goudsmit's proclamation, Ingelfinger's was met with immediate vocal criticism from journalists, largely because they feared that it would hamper journalists in their coverage of medical research conferences—for example, by making researchers reticent to clarify or extend their conference remarks to journalists. The editor of the NASW's newsletter declared that the Ingelfinger Rule "amounts to censorship," in part because much medical research is publicly funded and therefore should be open to public scrutiny. Other NASW members raised a din in letters to the editor. But Ingelfinger remained steadfast, estimating in 1972 that he refused to publish six to seven accepted papers every year because of violations of the rule. He argued that journalists were misguided in their attempt to cover medical research conferences, and in their demand for access to the texts of papers presented at the meetings, because the papers frequently are extensively reworked after the conferences, as the researchers publish them in medical journals. "Although [journalists] pride themselves on reporting accurately, there is no assurance that what they report is accurate in the first place," Ingelfinger said.[51] So influential was the *New England Journal of Medicine* that the Ingelfinger Rule immediately began to cast a pall over exchanges with reporters at scientific conferences. Scientists began to be reluctant to provide the texts of their papers to journalists, even when they had no plans to publish their work in the *New England Journal of Medicine*, and journalists began to complain that conferences consequently were more difficult to cover and a less fruitful source of news—complaints that persist to the present.

However, Philip Abelson, the editor of *Science*, said in 1972 that his journal did not have a rigid ban against prepublication publicity. This claim is interesting because, in later decades, *Science* would embrace the Ingelfinger Rule with a vengeance.[52]

The journalists' dispute with Ingelfinger would continue through the 1970s and persist with his successors, Arnold Relman in the 1970s and 1980s, with an editor at the *New York Times* threatening to use the federal Freedom of Information Act to obtain copies of reports on federally financed research despite the Ingelfinger Rule, and Marcia Angell and Jerome Kassirer in the 1990s.[53]

Widening Leaks: The 1970s and 1980s

The decades of the 1970s and the 1980s saw an increase in the sparring between journalists and journal editors over the control of news about science and medicine. Much of that conflict continued to center on the Ingelfinger Rule, with its proponents seeking to expand it internationally, as part of a move toward voluntary global publishing standards for medical journals advocated by an ad hoc group called the International Committee of Medical Journal Editors. Stephen Lock, one of the participants, writes that the process began after "a medical secretary in Seattle wrote to the editors of several major American journals complaining of having to retype articles each time they were rejected for publication because of the different reference style each of them used."[54] A medical conference in Vancouver, British Columbia, in January 1978 provided a convenient venue for editors of several medical journals to meet to develop joint standards for reference standards. The International Committee of Medical Journal Editors—colloquially known as the "Vancouver group" because of the site of its first meeting—published its first edition of standards for biomedical research in three journals in 1979, and several later editions, to which more than five hundred journals have agreed to adhere. The first edition dealt purely with matters of bibliographic style, but by the third edition, the guidelines included a section on "prior and duplicate publication" that neither defines nor explicitly forbids such practices, apparently leaving each journal the latitude to unilaterally determine its own policies in this area. The most recent edition, issued in 2003, endorses Ingelfinger's approach: "Researchers who present their work at a scientific meeting should feel free to discuss their presentations with reporters, but they should be discouraged from offering more detail about their study than was presented in their talk." The 2003 statement also endorses the use of a news embargo:

> Editors can foster the orderly transmission of medical information from researchers, through peer-reviewed journals, to the public. This can be accomplished by an agreement with authors that they will not publicize their work while their manuscript is under consideration or awaiting publication and an agreement with the media that they will not release stories before publication in the journal, in return for which the journal will cooperate with them in preparing accurate stories.[55]

The statement's equivocation about Ingelfinger restrictions at first seems odd, given the fact that Arnold Relman, Ingelfinger's successor at the *New England Journal of Medicine,* was one of the prime movers behind the international committee. Relman is an outspoken supporter of the Ingelfinger Rule and in fact was criticized during his tenure for allegedly trying to tighten its implications for press interactions.[56] The editors' vague statement on prior publication may be due to the fact that George D. Lundberg, then editor of the *Journal of the American Medical Association,* which had criticized the Ingelfinger Rule, also was a member of the international committee. However, Ingelfinger himself was unable to convince the Council of Biology Editors, another group of journal editors, to endorse his rule. "Our function, as I see it, is to enable scientists to report fully on their research from their own points of view—not a reporter's. Personally, I'm not too concerned about the medical press putting journals out of business," Edward Huth, editor of the *Annals of Internal Medicine,* said at the time.[57]

And officials at the *Journal of the American Medical Association*—in sharp contrast to its position under Morris Fishbein decades earlier—also indicated that they would not go along with Ingelfinger. "The AMA does not feel that press coverage interferes with formal publication at a later date," Ernest B. Howard, the association's executive vice president, told the National Association of Science Writers in 1973. "The AMA feels that press coverage helps bring the report to the attention of the medical profession much faster than through regular journal channels. The press report serves to draw more attention to the formal paper when it finally is published."[58]

Nevertheless, science and medical journalists continued to rely heavily on journals and press conferences, rather than their own enterprise as reporters, as sources of news. For example, *Science* also continued to get major press attention, presumably at least in part because of its embargo. A clipping service counted 252 articles citing the journal between June and September 1970. In 1976, the AAAS estimated that 75 science reporters were relying on the journal, but the association later estimated that during the year ending April 30, 1979, more than 400 newspapers, 20 general-interest magazines, and 50 specialized magazines and newsletters had reported on *Science*—with coverage appearing on 283 of the 365 days in the period.[59]

The journalists' continued reliance on prepackaged news in this era was illustrated by an analysis of the medical coverage of the *Detroit Free Press, Milwaukee Journal, San Francisco Examiner,* and *Washington Post* from 1967 to 1978. The study found that 11 percent of the newspaper articles cited a report at a medical meeting as their sole source, while 7 percent relied only on a scientific or medical journal. An additional 7 percent cited only a press release as their source. Another analysis, of 1983 medical coverage in Pittsburgh's two major dailies, found that 39 percent of stories were based solely on a press conference, and an additional 5 percent quoted only a journal article as a source.[60]

On the contentious issue of independent reporting of embargoed news, by the 1970s, it was well understood by journalists and public relations officials that journalists were not bound by embargoes on scientific studies if they had obtained the same information from nonembargoed channels. For example, the Associated Press ran a story in September 1971 about infections caused by some soft contact lenses, even though an embargoed release had been distributed about a paper on the problem scheduled to be delivered by an ophthalmologist at a conference later that month. The AP reporter learned about the problem and documented it through the California government. "Under the circumstances I didn't feel obligated by the release date because the release didn't initiate my story nor did I borrow from it," the reporter argued.[61]

When a journalist did acquire such information independently, a public relations official could sometimes persuade the journalist to hold the story anyway, but the cooperative journalist then ran the risk that yet another journalist would also stumble across the story and score a scoop of his or her own. That is exactly what happened in 1974, when news leaked out about the discovery of a subatomic particle called the J particle.

Walter Sullivan of the *New York Times,* who had made a specialty out of covering high-energy physics, got wind of the discovery by physicist Victor Weisskopf, who did not describe the nature of the discovery but asked him to delay publishing any news of it until it was published in *Physical Review Letters*. Shortly afterward, a source tipped Sullivan to the details of the discovery. Using a telephone in the press box at Yale's football stadium—where he was watching the Yale-Princeton football game—he got physicists at Stanford to describe the discovery. By halftime, George Trigg, one of the editors of the physics journal, authorized Sullivan to write the story, which appeared on the front page.[62]

Meanwhile, the American Institute of Physics was planning a press conference to announce the discovery. The day before the press conference, Earl Ubell, science editor of the *New York Herald Tribune,* notified Audrey Likely, the head of public relations for the American Institute of Physics, that he had independently received information on the discovery and would report it the next day, ahead of the scheduled announcement. As Likely recalled:

> He went on to say that he didn't feel bound by our restrictions, as he had already obtained the data on his own. I begged, pleaded, and cajoled that he not break the story ahead of time. I reminded him that he would have an extra three days to fine-tune his article and could therefore write in greater depth. I convinced him. He agreed to wait. Then, late afternoon the day before the big day, Harry Schmeck of the *Times* called to say that the *Times* syndicate in London had the J story on the wire. Jonathan Piel, then the public relations department's science writer, and I spent the next five or six hours calling AP, UPI, (putting the story on their wires) and every science writer we could reach to cancel the embargo and give them the data. I had the privilege of calling Earl. When I finished telling him, there was no verbal abuse, no screams, no curses. Just a great quiet, then a sigh, and a voice saying. "Leo Durocher said it all: 'Nice guys finish last.'"[63]

But as medical journals moved to expand their control over interactions between journalists and scientists, Goudsmit's physics journal relinquished that power. When James A. Krumhansl took over editorship of *Physical Review Letters* in 1975, he reiterated Goudsmit's policy against prepublication publicity in an editorial. "We again conclude that our profession and the public are not responsibly served when news releases preempt the critical viewing by the physics community." This declaration drew protests from several senior physicists—but apparently none from journalists. For example, Edwin Goldhasser of the Fermilab particle physics center wrote to the American Physical Society's president to complain that Goudsmit's rule hampered news coverage of scientific discoveries. "Why, I continue to wonder, do we continue to deprive ourselves of the healthy interest and participation of press and public at times of excitement in our field?"[64]

From retirement, Goudsmit weighed in against any change. He submitted to the society's governing board a letter endorsing Goudsmit's rule from Walter Sullivan, science editor of the *New York Times*. Sullivan argued that journalists are not qualified to assess the validity of scientific research and so should wait to report on research until it has passed through peer review. He concluded that "it seems to me that there is no question ... [that] the physics community is best served by your present restraints."[65]

Nonetheless, the physics society's governing council voted to void Goudsmit's rule, stressing "its dedication to the rapid and accurate dissemination of information to both the scientific community and the public. Accordingly, reports of new developments in the public news media will not be considered as prior publication prejudicial to acceptance of articles." Krumhansl then published another editorial in *Physical Review Letters* in which he offered rapid publication to exceptional research by omitting the usual peer review, so long as a senior scientist unconnected with the research endorsed the request for expedited publication. "We hope that the pressure for anticipatory mass-media publicity will thereby be reduced."[66] (The American Physical Society today continues to eschew embargoes on its journals.)

An embargo was a key feature of press coverage of a historic scientific conference in February 1975, at which molecular biologists debated rules for the performance of recombinant DNA research. In 1973, scientists had developed technology for moving genes from one organism into another. Some researchers began to fear that, left unregulated, such research could unwittingly produce a monster pathogen, and leaders in the field called for a moratorium on recombinant DNA research. Scientists and ethicists gathered at the Asilomar Conference Grounds in Pacific Grove, California, under the auspices of the National Academy of Sciences to attempt to draw up guidelines for conducting recombinant DNA research that would minimize the risk from the experiments. From the outset, planners envisioned inviting eight reporters to cover the meeting, on the condition that they would agree not to file any reports on the meeting until after its conclusion. Among the reasons that the scientists listed for limiting the number of reporters was that the conference center had limited seating. However, faced with the threat of a suit by the *Washington Post* for access to cover the meeting, Howard Lewis, the academy's press officer, convinced the organizers to expand the reporters' quota, first to twelve spaces and eventually to sixteen. The media outlets represented were *Bioscience*, the *Boston Globe*, *Chemical and Engineering News*, *Frankfurter Allgemeine*, the *Journal of the American Medical Association*, the *Los Angeles Times*, *Nature*, *New Scientist*, the *New York Times*, *Rolling Stone*, the *San Francisco Chronicle*, *Science*, *Science News*, the *Wall Street Journal*, the *Washington Post*, and the *Washington Star-News*. No wire service or broadcast reporters were invited; broadcast journalists were intentionally excluded out of a belief that the exchanges among scientists at the conference would be dampened if they knew they were being recorded for broadcast.[67]

Each of the sixteen reporters—as well as several others who showed up, uninvited, at the meeting but were permitted by Lewis to stay—agreed that they would not publish any stories until after the conference ended. There was a briefing for reporters every afternoon, and the journalists were permitted to interview participants freely. The journalists were also permitted to tape the proceedings, although conference attendees were jittery about this, and on the first day nearly voted to ban the use of tape recorders.

But one enterprising journalist nearly wrecked the embargo arrangement. The Associated Press, which did not have a reporter of its own at the Asilomar meeting, inquired whether the *Monterey Peninsula Herald*, a local afternoon newspaper, was planning to cover the conference. The paper's managing editor was unaware of the meeting, and he dispatched one of his reporters, Kevin Howe, to file a report. After checking with the manager of the Asilomar facility, Howe stopped by the meeting on its opening day. He immediately encountered a French reporter, newly arrived at the meeting, who was loudly complaining about the embargo. "Howe had been a reporter long enough to know that there are times to ask questions and times not to ask questions," one account of the Asilomar meeting recalls. Howe instead picked up copies of various scientific papers, tried to digest them on his own apparently without speaking with any of the scientists or other reporters, and returned to the newsroom to write his piece, which appeared that afternoon.[68]

The seven-paragraph story, though sketchy, appears to have been accurate. Headlined "Scientists Study Risks of New Organisms," the unbylined piece began: "A conference of scientists opened at Asilomar this morning to consider the possible danger of producing new and harmful bacterial strains—some toxic, others disease-carrying, and still others immune to modern antibiotics—through laboratory experiments." The article suggested that a major point of debate at the conference would be a National Academy of Sciences report that proposed technical safeguards that would be intended to prevent accidental release of genetically engineered organisms.[69]

The Associated Press picked up a version of Howe's article and sent it out to other newspapers that afternoon. The *San Francisco Examiner,* which did not have a reporter at the meeting, published it. Lewis later recalled that he was confronted by one of the scientists at the conference: "'Lewis,' he growled, 'so much for trusting reporters. They broke your #@&*$+% embargo!'"[70]

Howe's article also caused problems for the reporters covering the meeting when their editors saw the AP story. Several editors called their reporters at Asilomar to demand an explanation. All of the reporters apparently convinced their editors not to run the story and instead to wait until the conclusion of the conference so the reporters could file their own coverage. David Perlman, then the science writer for the *San Francisco Chronicle,* later recalled that he told his city editor: "Don't use this AP story. It doesn't mean anything. It's in the middle of a session. They haven't really decided anything." The city editor went along. "I'm the science writer so, he believed me. It was just that simple," Perlman said. Despite the AP article, the reporters at the conference agreed to continue to abide by the embargo. When a reporter from United Press International showed up at the conference—perhaps in response to the AP article—Perlman convinced him to adhere to it as well.[71]

The Associated Press asked the *Monterey Peninsula Herald* for more coverage of the Asilomar meeting, and so, the next day, Howe returned to the conference. He encountered reporters complaining about the AP story and accusing one another of being the source—and eventually deduced that his story was the source of their consternation. Eventually, he met Howard Lewis, who explained the embargo rule to him. Nevertheless, Howe did file another piece, which described the potential risk from DNA research and the proposed biohazard rules in greater detail. He extensively quoted Mortimer P. Starr, a professor of bacteriology at the University of California at Davis, as questioning whether the rules were needed or would be effective. "We have to ask . . . what would the costs be?" Starr was quoted as saying. "Is it worth it? Do we stop the work, or do we let others do it who aren't as concerned about this issue?"[72]

Perlman not only protected himself from competition by convincing the UPI reporter to honor the embargo; he also manipulated the embargo to his competitive advantage. The participating reporters had agreed that the embargo would lift when the meeting ended, and that was scheduled for a Thursday afternoon. Perlman convinced Lewis to add a closing press conference after the last scientific session. The *Chronicle* is a morning newspaper, so the earliest that Perlman's story could get into print would be Friday morning. His real motivation in requesting the press conference was to forestall the United Press International reporter from filing a story long enough so that the *San Francisco Examiner,* the *Chronicle*'s main competitor and an afternoon paper, would not receive the wire story in time to publish it in its Thursday edition.[73]

Coverage by the newspaper journalists at Asilomar was largely reassuring in tone. The stories noted the risks posed by genetic engineering research but stressed that the scientists at the meeting had agreed both to follow unprecedented security measures to prevent outbreaks and to eschew some experiments that they deemed too risky. Statements from critics were rare; when the reporters did quote critics, those quotes appeared near the end of the stories. The stories did mention the controversies and debates that had occurred throughout the meeting, but by focusing on the conference's report, the news stories generally conveyed an air of scientific consensus rather than one of scientific controversy.[74] (None of the three major broadcast television networks covered the Asilomar report at all.)[75] One scientist characterized the coverage thus: "The members of the press who had attended throughout were now freed from their bonds; and having earned honorary degrees in molecular biology, they released generally laudatory and respectful commentary."[76]

Cristine Russell, who covered Asilomar for the monthly magazine *Bioscience* and who later was a science writer for the *Washington Star* and the *Washington Post,* said twenty-five years later that daily news coverage of the conference, in the absence of the embargo, would have been confused, because the scientists' deliberations were themselves somewhat discombobulated. "It took us a while to get to know what they were talking about and also to get them to talk to us. . . . [In] the first two days, if we had tried to write stories, they would have been fractured stories that would have not really given a sense of what was evolving," she said. "The stories that resulted were far superior to those that might have been written without an embargo." The embargo also offered journalists a chance to take time to understand recombinant DNA technology and its risks before writing about them, she said. "Perhaps our coverage was more positive than it might have been if we had started off in an antagonistic fashion, but I think we had balanced coverage because we learned a lot more [and] were better educated. We really did understand how the research was done. And we developed sources at that meeting that I think are still sources for many of us today 25 years later."[77]

Journalists' increasingly uneven adherence to embargoes and scientists' increasing propensity to announce results to the public rather than to peers led Edward Edelson of the *New York Daily News* to declare in late 1980 that the "old principle of scientific communication is collapsing." Although the embargo system supposedly ensures that scientific accounts of research are not preempted by popular ones, "the net result is that most scientists hear about a finding in the newspapers or on television." He suggested that the embargo system be replaced with a "free-for-all. . . . We have that system now in every other field; it is called the free exchange of ideas."[78]

At about this time, science journalists also found themselves in the unaccustomed position of being the subject of research themselves, when communication researcher Sharon Dunwoody published studies documenting cooperative behaviors by science journalists covering the AAAS annual meeting in 1977 in Denver. Her research did not address embargoes, but it did cite other cooperative behaviors such as supportive questioning at news conferences and sharing of expertise among reporters. These findings apparently caused discomfort among some science journalists: after Dunwoody's research was published, unidentified science journalists told her that their news-gathering practices at conferences differed from their news-gathering behaviors at other times.[79] This suggests that the journalists did not draw the obvious parallels between Dunwoody's findings and their heavy, routine reliance on embargoed materials.

However, one veteran science reporter, Jerry Bishop of the *Wall Street Journal*, argued that Dunwoody erred in her conclusion that the AAAS uses press conferences to control science reporters' coverage of its conferences. Rather, Bishop wrote, press conferences at the meetings developed at the behest of journalists, as shown by journalists' reaction when no press conferences are scheduled: "There is an air of suppressed panic as the reporters paw through the meeting program trying to figure out which session might offer news. . . . Meetings without press conferences makes science reporters nervous and anxious and they don't like to be nervous and anxious. Perish the thought that one might file an exclusive story; after all, how would the desk know it was a good story if the *New York Times* and the AP didn't carry it also."[80]

Even when enterprising reporters gathered research news independently, they could run afoul of the embargo system. In 1986, the *Journal of the American Medical Association* dropped the *Miami Herald* from embargoed access after accusing the newspaper of prematurely reporting an AIDS study. The reporter claimed that he had researched the AIDS story without relying on embargoed materials; free of the embargo, he independently reported several other studies from the journal in advance of the embargo release time.[81]

Another long-running controversy that began to take on new energy centered on whether the Ingelfinger Rule wrongly delays the public from finding out about potentially life-saving medical treatments. This issue erupted forcefully in January 1988, when the *New England Journal of Medicine* published preliminary results of a study that suggested that taking one aspirin tablet every other day can lower a person's chances of suffering a heart attack. The preliminary results were so convincing that, on December 18, 1987, the researchers concluded that they were ethically bound to end the experiment early, and they approached the journal about the prospects for quickly publishing the results. The journal and researchers agreed that the results would be published in the January 28 issue, and the researchers prepared letters to the 22,071 physicians who had participated in the study, summarizing the results and informing each participant whether they had been taking aspirin or a placebo. Mailing of the letters was timed to coincide with the journal's publication.[82]

But the Reuters news service broke the embargo a day early and carried a report on the study, triggering a spate of stories by other news media. The leader of the research project criticized the embargo failure, saying "premature release of the results occurred in a haphazard manner and led to considerable confusion." One scholar subsequently searched for errors of omission, inference, and sensationalism in the stories that were hurriedly broadcast and published, and he found examples of each type of error. For example, the five newspapers committed an average of 5.6 errors of omission, by failing to report details such as the age range of the participants in the study. Errors of sensationalism included characterizing the study as dramatic and unexpected and mischaracterizing the researchers' decision to end the study three years early. Errors of inference included generalizing the results of the study, which involved male physicians of a specific age range who had no previous heart disease or strokes, to other populations, such as all men or the general public.[83]

Lawrence K. Altman, a physician and medical writer for the *New York Times*, wrote a front-page story in which he asserted that results of the study had been known for weeks among an inner circle of "scientists, government officials, business leaders, and journalists." More disturbingly, Altman charged that the journal's Ingelfinger Rule delayed the dissemination of information that could have had a major impact on public health.[84]

Others agreed. In a letter to the editor published in the *Times*, one researcher claimed that "thousands of the heart attacks that occurred in American men between Dec. 18 and Dec. 28 could have been prevented" if the journal had permitted the researchers to announce their findings before the journal's publication, although one of the journal's editors subsequently disputed that conclusion.[85]

In punishment for its embargo violation, the editor of the journal,

Arnold Relman, suspended Reuters's embargoed access to the journal for six months—even though the news service's executive editor claimed that Reuters learned about the study independently. Relman also warned that he would not tolerate the wire service obtaining advance copies of the journal from others during those six months. "If we find that Reuters tries to obtain copies during the suspension we're prepared to put them on the blacklist permanently," he told the Associated Press. "And if we find out who was supplying them the copies, we will take similar action."[86]

For its part, Reuters seemed unfazed. Its North American executive editor said that Reuters intended to "obtain copies of the journal and issue stories on news merit." He said he would not abide by the embargo release time. "Why should we? . . . We don't have an agreement, do we?" Reuters then proceeded to break stories about research published in the journal on topics including cirrhosis of the liver, nonsurgical treatments for enlarged prostates, and drug resistance in viruses. The *Boston Globe* concluded that the wire service "now routinely jumps the NEJM embargo."[87]

The impasse continued for more than a year. A key sticking point was Reuters's assertion that it should be able to ignore the embargo if the news service concluded that embargoed information—such as the results of a drug trial—had somehow become known to Wall Street investors and was influencing the market, even though the embargo had not yet expired. "Everybody has a right to invest in the stock market and no one should have an advantage over any other one," said Andrew Nibley, the newly named editor of Reuters's American operation. "Wall Street has copies of the New England Journal. Wall Street analysts have it in their hands a day before anyone else does. And it's not fair. We cannot withhold information when we see it's moving share prices or is likely to move share prices. It's not fair to all investors." In March 1989, the journal reinstated the wire service's embargoed subscription, although the two organizations continued to disagree about embargoes on market-sensitive information. And the wire service did continue to break stories ahead of the embargo, such as a 1990 article on strokes in young people who smoked crack cocaine.[88]

The Reuters incident appears to have triggered some introspection among participants in the embargo system. In an editorial, *Nature* said that Relman had been "heavy-handed" toward Reuters and challenged his assertion that the embargo provides physicians a chance to read about medical research before patients are informed about them in the news media and start inquiring. "Are physicians that diligent, or the mails that good?" The editorial also acknowledged that the embargo system gives journals "the benefit that their contents are likely to be noticed in writing by more periodicals than would otherwise have done so." Meanwhile, Daniel Greenberg, a veteran science journalist, concluded that the embargo system reveals that science journalists as a group are lazy. "If the press wants speedy access to new research findings, it could pursue scientific information with the same vigor that it goes after political scandal. But if it wants to be spoon fed, it should abide by the rules of the feeders."[89]

But if others criticized Relman, subscribers to his journal did not. The journal mailed surveys on the issue to a random sample of 1,607 of its U.S. subscribers. Seventy-four percent responded, and of them, 86 percent said that they wanted the embargo to continue. "We take that response to be a strong mandate for the continuation of our present course," Relman wrote.[90]

And, some said, the incident exposed the fragility of the embargo arrangement. "If Reuters decides not to abide by the rules and other news organizations follow suit, Relman's hard-nosed stance could result in the collapse of the embargo policy," an article in *Science* observed. The incident "could dramatically change the way medical news is reported," said the *New York Times*.[91]

Embargoes also shaped the reporting of the March 1989 announcement by researchers at the University of Utah that they had discovered a process of "cold fusion," or fusion at room temperatures. University administrators at first would not permit the university's press office to distribute an embargoed announcement describing the discovery, because the administrators believed that the embargo would be violated. But the head of the university's press office insisted that the university would have to provide some embargoed information about the discovery in order to convince national news reporters to travel to Utah for a press conference discussing the discovery. The university administration permitted the news office to share only a one-paragraph announcement with journalists in advance of the press conference. Consequently, it appeared that few national journalists would report on the experiment. However, Martin Fleischmann, one of the researchers, spoke freely with a reporter for the *Financial Times* in London about a week before the scheduled press conference. The story was distributed by wire services the day before the press conference, triggering wide coverage. Fleischmann later said that he did not realize that his interview voided the news embargo.[92]

Later in 1989, another media feeding frenzy developed when the Reuters news service reported in August that researchers had discovered the gene that causes cystic fibrosis. Papers describing the discovery were not scheduled to be published in *Science* until the following month; technically, Reuters's reporting was not a violation of the news embargo because the papers had not yet been fully edited, much less distributed to journalists under the embargo. But Daniel Koshland, the editor of *Science,* decided to allow an exception to the Ingelfinger Rule and allow the researchers to discuss their findings with reporters. "Once Reuters had broken the story we thought it was unfair to the rest of the press to withhold the information," he said. Using private jets supplied by the Howard Hughes Medical Institute, the researchers held press conferences in Toronto and Washington, D.C., on the same day. At the same time, patent officials at the University of Michigan, home base of one of the key researchers, became concerned that the prepublication publicity might endanger the university's prospects for patenting the potentially lucrative discovery, so they raced to file a patent application at 11:45 P.M. on the day of Reuters's story.[93]

Despite the conflicts over the terms of the embargo and their interpretation, science journalists in this period still endorsed embargoes overwhelmingly. In a survey of its members by the National Association of Science Writers in 1989, 84 percent of respondents said that embargoes were justified under certain circumstances; when presented with a list of possible justifications for embargoes, 53 percent agreed with giving journalists an equal chance at the story. One-quarter said that they would break an embargo if they thought someone else would, and 20 percent said they would break an embargo on a story that was "too important to hold" (see table 2.1.).[94]

And the journals continued to defend embargoes. Relman, the editor of the *New England Journal of Medicine,* described his rationale: "The medical profession should hear about the results of research at least as soon as the public hears about it. . . . When the public hears about it, physicians should have their journals on their desk, or in the library so that they may consult the journal. They will then have the information to respond when patients call and ask, 'doctor what about this treatment or this operation?'"[95]

The Ingelfinger Rule also continued to cause consternation and confusion among scientists. In October 1988, the National Institutes of Health's Recombinant DNA Advisory Committee was asked to grant permission for the first experiment in which a new gene would be transplanted into human beings. But the researchers in charge of the experiment declined to provide the committee with important raw data, including data about the safety of the proposed experiment, out of fear that the data then would become part of the public record. That would make the data available to competitors and could also interfere with publishing the data in the *New England Journal of Medicine* or *Science,* the researchers told the committee. The committee

Table 2.1 Embargo-Related Views in a 1989 Survey of Science Journalists

Survey Question	%
What reasons would justify the embargo of journal articles?	
Not under any condition	24
To give everybody an equal chance at a story	34
To give the scientific community time to see the article	26
To coincide with the date the article is published	39
To ensure consistency and accuracy of coverage by the press	30
What reasons would justify the embargo of news releases?	
Not under any condition	18
To give everybody an equal chance at a story	46
To give the scientific community time to validate scientific results	29
To avoid irresponsible reporting of scientific results	35
To ensure consistency and accuracy of coverage by the press	33
Under what conditions would you break an embargo?	
If I have reliable information that the embargo will be broken by someone else	18
If someone else has broken the embargo	51
If the story is too important to hold	14
If I feel that withholding the story will compromise public welfare	52
For other reasons	8

Note: N = 393. Responses are from survey participants who indicated that they were "science journalist[s] with no public information responsibilities." Source: National Association of Science Writers. Tabulations by the author.

unhappily endorsed the experiment, but the director of the NIH overruled them and delayed the experiment, declaring that the Recombinant DNA Advisory Committee "will not be held hostage to the *New England Journal of Medicine.*" The editors of the two journals insisted that the disclosure would not have hurt their chances for publication because the journals permit researchers to provide data requested by governmental bodies, and the experiment subsequently proceeded.[96]

New Technology in the Midst of Mounting Controversy: The 1990s and Beyond

The 1990s opened with a gambit by the *Journal of the American Medical Association* to fine-tune the embargo to its advantage: in April of that year, the journal changed its publication date from Friday to Wednesday, making it a day before the release date for the *New England Journal of Medicine* rather than a day after. "If two of the principal medical journals in the world are working on similar research, we want to be the first out on it," a spokesman for the American Medical Association said.[97]

It worked: coverage of the *Journal of the American Medical Association* by the *New York Times* jumped by 50 percent after the embargo time was made earlier in the week in 1990. The paper's coverage of the *New England Journal of Medicine*—the more elite of the two journals—remained unchanged. This suggests that the New England journal had overshadowed the AMA journal when the New England journal had an earlier embargo release time; when the AMA journal made its release time earlier, it emerged from the shadows.[98]

The importance of the two medical journals as news sources in this era was underscored by a 1991 survey of 262 U.S. medical journalists. Asked to identify their news sources in a normal one-month period, 34 percent cited the *New England Journal of Medicine* as a source that was used "often," and 33 percent placed the *Journal of the American Medical Association* in that category. *Science* was reported as used often by 25 percent, and the *Lancet* was reported as used often by 16 percent. By contrast, press releases were used often by 10 percent, research papers by 60 percent, and personal contacts in the medical community by 58 percent.[99]

Broadcasters and print journalists continued to tussle over the embargo as well, as evidenced by a dispute over who was responsible for an embargo violation in July 1995. *Nature* had disclosed to journalists, under embargo, the text of a paper describing a genetic cause for one form of Alzheimer's disease. But about twenty-four hours before the embargo expired, the paper was mentioned publicly during a congressional hearing on the importance of biomedical research: "At 6 tomorrow an embargo will be lifted to talk about a new gene that is the major gene for early Alzheimer's Disease," Allen D. Roses, the head of the research team, told the Senate Special Committee on Aging. "On Thursday it will be published in the journal *Nature*. With that gene . . . we cover the genetic field of Alzheimer's Disease by about 95 percent of the prevalent cases." Roses provided no details other than that the gene would be designated S-182, and the committee members did not question him. That evening, *ABC World News Tonight* reported Roses's statement but disclosed no additional details from the *Nature* paper. "There was very major news almost made here in Washington today about Alzheimer's disease," anchorman Peter Jennings said while introducing a report on the hearing. "Researchers writing [in] the magazine *Nature* will report in full tomorrow that they have isolated the gene that appears to cause Alzheimer's in people who are under 60. Alzheimer's is ultimately fatal and slowly destroys the brain. Today, the head of the panel that conducted the research would only tell a Senate committee that much."[100]

Judging that ABC News had violated the embargo and had consequently freed other journalists to write about the project, *USA Today* published an article the next morning with details from the *Nature* piece. John Maddox, *Nature*'s editor, called *USA Today* "disingenuous," noting that *Nature* had mentioned plans for the paper in the front matter of its previous issue. (However, the mention was cryptic at best: a small box in the previous issue's table of contents listed several topics to be included in the journal's next issue; included in the list was "Alzheimer's mutation.") Maddox banned the newspaper from embargoed access for six months. "While it may seem to them they are being penalized unfairly for breaking bureaucratic rules, their real disservice is to their fellow journalists," Maddox wrote. Susan Weiss, the newspaper's managing editor, replied to Maddox that ABC's broadcast violated the embargo. "Your willingness to allow Peter Jennings to report the discovery . . . raises again the concern of favoritism by medical and science journals toward television networks. The embargo policy itself already discriminates against print media by giving scientific and medical news to network broadcasters a day ahead of newspapers."[101]

In the 1990s, the growing commercial importance of biotechnology undermined the strength of embargoes on research in the field, while companies sought ways to gain access to embargoed information and turn it to their benefit. For example, in 1996, the *New England Journal of Medicine* published a study that determined that the antiobesity drug Redux would cause a lung disorder in some people. The journal also published an editorial that concluded that the drug would save more lives than the deaths it would cause. American Home Products Corporation, which markets the drug, got an advance copy of the editorial; the source, the company said, was the French pharmaceutical company that sponsored the study. Before the embargo lifted, the company issued an unembargoed press release touting the journal's positive editorial, in an apparent effort to draw attention from the negative conclusion of the study. "The NEJM findings represent information we have already looked at very carefully and shared publicly," the press release said. "This paper does not represent a new study or new data." The stock of the drug's manufacturer rose, apparently because of investors' anticipation that the editorial's conclusions would promote increased sales of the drug, although the research subsequently became embroiled in controversy over an apparent financial conflict of interest by the editorial's authors.[102]

The continuing iron grip of the Ingelfinger Rule was illustrated by one incident at the 1994 annual meeting of the American Association for the Advancement of Science, the publisher of *Science,* in San Francisco. Although the AAAS meeting had long developed a reputation among science journalists as being thin on news, the 1994 meeting featured one nugget of interest to reporters: the presentation of results of a controversial experiment to combat global warming by depositing iron filings in the ocean, which would promote the growth of phytoplankton, which would remove carbon dioxide from the air through the process of photosynthesis. Despite journalists' interest in the research, the paper on the experiment was not made available to journalists, and the researcher who led the experiment did not attend a press conference organized by the AAAS to discuss the main points of the conference session. The reason was that the researcher had submitted a paper to *Science,* and such prepublication dissemination of the data could jeopardize his chances of publishing the paper. An irate David Baron, a science reporter for Boston's WBUR-FM, wrote in an open letter to the AAAS: "I now understand that your public affairs staff has been instructed to hide the most newsworthy studies from the media, and that the news conferences are, by design, where only old results will be discussed. . . . I thought the AAAS was committed to the advancement of science; I now realize it's committed only to the advancement of *Science.*" In response, the AAAS's director of communications said that her staff had an obligation to warn speakers who planned to submit papers to journals that follow the Ingelfinger Rule about the risk posed by appearing at a press conference at the session. But, she noted, the Ingelfinger Rule restricts researchers, not journalists. "Thus, any reporter is welcome—and encouraged—to comb the AAAS program for sessions of interest."[103]

Embargoes also figured in one of the most important research developments of the decade—the February 1997 announcement by Scottish researchers that they had cloned a sheep named Dolly. The research report was published in *Nature,* which had followed its usual practice of distributing advance embargoed copies of the paper to journalists. But because of the importance of the finding, many reporters expected that the embargo would be violated. For example, Gina Kolata of the *New York Times* and her editor decided that the embargo would probably fail. "We decided that I would get a major story ready to go and that the *Times*'s editors would . . . alert us immediately if another news organization had reported *Nature*'s cloning story. If and when that happened, the *Times* would rush my story into print."[104]

Indeed, the embargo did bust, but not because of a journalist with embargoed access. The Italian news service ANSA first carried news of the research during the weekend after *Nature* distributed the embargoed information to journalists, and three Italian newspapers then ran stories. Robin McKie of the London *Observer* also published a story, based on independent reporting, on the Sunday before the scheduled release time. "Scientists have created the first clone of an adult animal. They have taken a cell from a sheep's udder and turned it into a lamb," began McKie's story, which did not mention *Nature*. Kolata's foresight was rewarded. She later crowed: "*The New York Times* immediately went into action, publishing its story in time for the second edition of the paper, which meant that only those readers who lived closest to New York City and got the edition of the Sunday paper that was printed late Saturday afternoon could have been unaware of the cloning of Dolly."[105]

Thanks to the reach of the paper's own wire service, versions of Kolata's story appeared in other newspapers' Sunday editions as well, including the *Austin American-Statesman*, the *Dallas Morning News*, the *Dayton Daily News*, and the *Palm Beach Post*. The national syndication of Kolata's story appears to be the basis for a claim made on the dust jacket of her 1998 book on cloning that she "broke the story nationally." Such a claim dubiously tries to take credit for having a story ready to run in case another journalist's enterprise or scrupulousness led to an embargo break. Moreover, Kolata's claim notwithstanding, she and the *New York Times* were not the only ones to move quickly: the *Los Angeles Times* was also able to report the cloning in its Sunday editions. The front-page story summarized the research and raised the ethical issue of cloning humans. One reporter for the *Los Angeles Times* later concluded that the Dolly story "illustrates how important science news often is more a product of news management by the journals that publish peer-review research, than of any one reporter's special expertise or investigative energy."[106]

Another example of the vulnerability of embargoes to enterprising reporters was the dramatic 1996 announcement of the discovery of evidence of fossilized life in a meteorite from Mars. The researchers were set to publish their findings in *Science*, but the American Association for the Advancement of Science and the National Aeronautics and Space Administration—where most of the research for the project was conducted—jockeyed back and forth about how the information would be released, with each side seeking embargo arrangements that would better suit its ends and needs. As the publication date for the paper drew near, presidential adviser Dick Morris bragged about the meteorite discovery to Sherry Rowlands, a prostitute with whom he had a long-running relationship. Rowlands recorded in her diary that Morris boasted that he was one of seven people who knew about the existence of life on another planet. After the news of the discovery subsequently broke, Rowlands would sell her story to the *Star*, a supermarket tabloid. News that a key presidential adviser had discussed government business with a prostitute would force Morris to resign from his job.[107]

However, before embargoed copies of the paper even were distributed to science journalists, an enterprising journalist who was not part of the embargo system broke the story in a space-industry trade newspaper, relying on confidential sources and information he had gleaned from a scientific conference. His article triggered an immediate media feeding frenzy in which science reporters scrambled to describe the findings without the luxury of several days' worth of an embargo. The meteorite had been discovered by researchers funded by the National Science Foundation, but the National Aeronautics and Space Administration influenced press coverage to build support for its plans for Mars exploration. In fact, foundation officials were taken by surprise by the researchers' announcement, even though the research team included the head of the National Science Board, which oversees the foundation; that researcher, like all others seeking to publish in *Science,* was barred by the journal from disclosing the substance of the research until it was published. This aspect of the story illustrates the possibility of using embargoed information to influence political processes.[108]

In another example of enterprising reporting, the London *Observer* broke the story in 2001 that analysis of the human genome had found that humans have a surprisingly low number of genes—even though the story had been embargoed by both *Nature* and *Science* in an unusual shared embargo. The *Observer* had not agreed to the embargo, however, and its reporter obtained details on the research at a public biotechnology conference, three days before the embargo was scheduled to lift, at which a key scientist in the project spoke.[109]

But such incidents did little to weaken journalists' devotion to embargoes. For example, Boyce Rensberger of the *Washington Post* strongly praised embargoes in a science-writing manual published in 1997 under the sponsorship of the National Association of Science Writers. The embargo system, he wrote, "is a very good thing because science stories are more complex than ever and it takes time, sometimes several days, to do a good job. The embargo system removes the temptation to beat the competition, giving us more time to do our jobs well and giving the readers better-written stories."[110]

One vocal critic of embargoes was the *New York Times*'s medical reporter, Lawrence Altman, who continued to blast the Ingelfinger Rule and the embargo system in a two-part essay published in 1996 in the *Lancet,* a British medical journal that itself provides embargoed copies to journalists. Altman concluded that there is little evidence that the Ingelfinger Rule improves the quality of scientific journals and therefore should be dropped. Altman's argument led the *Lancet*'s editor to write that he was inclined to rescind its version of the Ingelfinger Rule. "Perhaps the question boils down simply to this: can editors trust investigators to report their research responsibly, and if not, why not?"[111]

Nevertheless, embargoes continued to expand in the 1990s. The British medical journal *BMJ* started distributing embargoed press releases to journalists, and noted that the press began quoting the journal more frequently.[112] In fact, the number of journalists that participated in science embargoes undoubtedly grew during the 1990s, due largely to the dramatic growth that decade of computerized communications. Even before the Internet became popular, the National Association of Science Writers began operating a section of the CompuServe computer service that was restricted to NASW members. Embargoed information from *Science, Nature,* and other journals was posted regularly on this service. This effectively placed the National Association of Science Writers in the position of enforcing the journals' embargo rules.

The growing popularity of the Internet in the mid-1990s provided a new route for dissemination of embargoed information. A commercial venture called Quadnet, started in the early 1990s, sent press releases from universities and companies to science reporters by electronic mail. Journalists participating in the system were warned that they would be dropped from the service if they violated embargoes on the electronic press releases. Another commercial venture, called Newswise, started distributing embargoed university press releases to journalists in 1991. In 1996, it moved to the World Wide Web, where approved journalists can use a password to retrieve embargoed press releases. Also in 1996, the American Association for the Advancement of Science started offering a site on the World Wide Web called EurekAlert! that journalists could use to obtain embargoed information from its journal as well as from other journals and universities that participated in the service. By late 1998, 1,993 reporters from 863 media organizations around the world were registered to use EurekAlert! and 278 institutions were supplying information for posting on the service. The service's advisory board at the time included journalists from National Public Radio, the *Los Angeles Times,* the *Chronicle of Higher Education, U.S. News & World Report, Business Week,* the *Washington Post, Science,* the *New York Times,* and the *Wall Street Journal.*[113] As noted in Chapter 1, in 1998 European science organizations countered EurekAlert! with AlphaGalileo, a Web site that disseminates news about scientific research conducted in Europe.

The immediate popularity among journalists of sites such as EurekAlert! and AlphaGalileo is evidence of the continuing power wielded by publishers of embargoed journals. Indeed, journalists signed up for these services explic-

itly in order to receive embargoed materials, not to challenge the authority of embargoes. However, the high visibility of these Web sites may also
have served to stimulate debate among some science and medical journalists
about the wisdom of embargoes. For example, in late 1998 *Science*—one of
the major beneficiaries of the embargo system—published a ten-page section of news articles that explored the benefits and drawbacks of journal
embargoes. And several months later, the National Association of Science
Writers held a morning-long workshop on embargoes that attracted several
hundred journalists.[114]

Despite the popularity of the Internet for disseminating embargoed materials, some public relations officials found themselves wary of the use of
e-mail and other electronic communications for disseminating embargoed
materials, particularly as more and more reporters from around the world
began to request to receive them. For example, Steve Maran, an astronomer
and press officer for the American Astronomical Society, said in 1996 that
he had stopped e-mailing embargoed details of scientific findings to be presented at the astronomical society's conferences. "There's a lot of people we
don't know, a lot of people, now in South America and throughout Europe,
on that list and in many different time zones," Maran said. Also, electronic
distribution creates new ways for the information to leak, he said: "I have
no doubt that when you broadcast a release, whether by e-mail the way I do
it, or by a password-protected Web page . . . that inevitably, unauthorized
parties gain access, not necessarily be surreptitious methods, but by, perhaps,
simply insufficient care on the parts of recipients."[115]

Some science journalists did use the Internet to make end runs around
the embargo system. For example, in 1995, a reporter from the *Chronicle of
Higher Education* used an Internet newsgroup devoted to particle physics
to gather information on the discovery of a new subatomic particle, called
the top quark, days before the news was formally announced. A graduate
student who had attended a seminar on the discovery posted details on the
newsgroup—which the reporter read and used as the basis of telephone
interviews with scientists. The reporter had all the information he needed
days before the *New York Times* and *Chicago Tribune* broke the story. However, because of his publication's weekly schedule, he was unable to break
the story first.[116]

Journal editors tried to prevent such incidents by construing the Ingelfinger Rule to forbid prepublication dissemination of scientific findings via
the Internet. Although scientists decried this tactic, journalists were largely
silent on the issue, perhaps because few yet use the Internet for news gather-

ing enough to understand the implications.[117] Alternatively, the journalists may have been satisfied with the current embargo arrangement and had little intention to use the Internet to subvert it.

Conclusions

Journal publishers and scientific societies today dominate the operation of embargoes on scientific information by deciding which journalists will have embargoed access, the terms under which that access is exercised, and punishment for violations of the embargo terms. However, this state of affairs evolved from a very different starting point, early in the last century, in which journalists—not scientists or their associations or publishers—insistently demanded advance access to conference papers and journal articles. The vendors of scientific information subsequently came to understand the power that embargoes gave them over science and medical journalists, and they have not been chary of exercising that power. Meanwhile, journalists by and large have continued to embrace the embargo enthusiastically, with the greatest source of complaints being fears that the terms of the embargo confer an advantage over one set of journalists, such as those working in television, over others, such as newspaper journalists.

3

Accuracy in Science Journalism

In June 1997, a team of Harvard researchers published a paper in the *New England Journal of Medicine* that examined the deaths of 2,051 nurses between 1976 and 1994. The researchers concluded that taking estrogen after menopause decreases a woman's chances of dying. Given the controversial nature of hormone-replacement therapy—which appears to protect a woman against heart disease but also may increase her chances for developing breast cancer—it is unsurprising that the paper received wide attention from the press.[1]

The *Washington Post* was one of the media outlets that covered the paper, which was routinely distributed to journalists in advance under the embargo. "Women who had been on [hormone-replacement therapy] for five to nine years had a 60 percent lower risk of death," the *Post* reported in a front-page story, bylined by David Brown, who himself is a physician. "Women who took supplements and who also had a sister or mother with breast cancer . . . had lower mortality (35 percent) than similar women not on hormone therapy."[2]

But, as the *Post* acknowledged in a correction the next day, both statements were in error. The risk of death was 40 percent lower, not 60 percent lower, for women who had used hormones for five to nine years. And the women with sisters or mothers with breast cancer and who had taken hormones had 35 percent less mortality than similar women not using hormones—not a mortality (or death rate from the disease) of 35 percent.[3]

These errors are understandable, in light of the complex statistical analysis used throughout the journal article. The article published in the journal had reported that women who had used hormones for five to nine years had a "relative risk" of 0.60 of death from all causes, compared to women who had never used hormones. Relative risk is an epidemiological statistic that compares the likelihood of contracting a disease by members of two groups.[4] In the hormone study, the relative risk of 0.60 meant that hormone users had 0.60 of the death rate of nonusers. The *difference* between the rates of the two groups (which is what Brown was reporting) is 1.00 minus that ratio, or 0.40.

Avoiding such errors is one of the principal arguments in favor of embargoes on scientific and medical journals. Scientists and physicians believe that such errors are common; in one recent set of surveys, physicians as a group said that reporters rarely are correct in reporting the technical details of medical or biological stories, while reporters disagreed with that assessment. Proponents of the embargo maintain that the arrangement gives reporters time to understand and report about complex topics, reducing the chances that they will commit errors in the process. Floyd Bloom, the former editor of *Science,* says that "the embargo period provides sufficient time for reporters to analyze and report on the often complex stories behind the data." Two top editors of the *Journal of the American Medical Association* have written about the use of embargoes: "We hope this approach helps contribute to accurate and orderly reporting about potentially complex medical research topics."[5]

This chapter will examine the relationship between the embargo and news reporting of science and medicine. First, we will consider the concept of accuracy—why accuracy is important, what scientists and journalists say it is, and how mass communication scholars have approached it. The chapter will review evidence about the level of accuracy in reporting about science and medicine, and it will consider the question of whether journal embargoes promote more accurate journalism.

The Imperative for Accuracy

Today, there is wide agreement that journalists in general, and science and medical journalists in particular, have a duty to be accurate in reporting, a responsibility that is grounded in the journalists' ethical imperative to tell the truth. Journalists themselves maintain that they have an ethical duty to be accurate. The ethics code of the Society of Professional Journalists, for example, calls on journalists to "test the accuracy of information from all sources and exercise care to avoid inadvertent error. Deliberate distortion is never permissible." The American Society of Newspaper Editors declares:

"Good faith with the reader is the foundation of good journalism. Every effort must be made to assure that the news content is accurate, free from bias and in context, and that all sides are presented fairly."[6]

Truth is a good in itself, and journalists should seek to be accurate in order to pursue that moral good. But accurate journalism is also an ethical imperative because of the effects that journalism has on society as a whole and on individuals within society. For example, the Hutchins Commission, in its seminal 1947 report that outlined the basic tenets of social responsibility theory, argued that accurate reporting was essential for democracy to operate well. "The first requirement is that the media should be accurate," declared the commission. Although the commission's view is common among journalists today, many journalists at first opposed the commission's argument that they have an ethical duty to provide accurate information in order to promote the effective functioning of democratic government. Today, most journalists agree that they have a responsibility to be accurate. However, many observers—and even some journalists—maintain that journalism has not fulfilled this ethical obligation.[7]

Accurate journalism about science and medicine is important because science and medicine play important roles in society. Consider medicine: reporting about the latest discoveries in medical research may serve the public interest by getting the word out about a new approach to treatment—or it may just as easily unadvisedly spur individual patients to stop taking medication without consulting their physicians. For example, one study found that news media coverage of a conference paper, reporting that one type of antihypertension drug increased a patient's risk of heart attack, led to a drop in the number of prescription claims for antihypertension medication of all types, with individual patients apparently deciding to stop using the medication without consulting their personal physicians.[8]

Media reports about research may also build political support for more government research funds for a scholarly field (and quite possibly as a consequence less research funds for some other field). This is equally true about many areas of nonmedical science, such as research into environmental risks (earthquakes, global warming, and the like). More fundamentally, leaders of the scientific establishment also maintain that journalism influences the public's understanding of how science operates and therefore can build (or erode) support for even highly abstruse pure science. Therefore, accurate journalism about science and medicine is important so that members of the public—and their political representatives—can form accurate judgments relating to science and research.[9]

It should also be noted that news organizations have a self-interest in seeking to be accurate. Some argue that the erosion in public trust of the news media can be tied to a perception of flawed reporting—both inadvertent errors and outright fabrications—and that accurate journalism could slow or reverse this erosion in public trust in journalism.[10] In addition, media organizations are increasingly confronted with a variety of legal challenges over their reporting. Accurate reporting can provide a legal defense against defamation lawsuits and other legal actions against a media organization.

However, accuracy is not the be-all for journalism. Upholding other values, such as speedy dissemination of news, can trump the value of accuracy in journalism. For example, journalists routinely publish or broadcast stories that are known to lack the explanation or response of a party in the story, even if it is possible that the party might point out an error or inconsistency in the story. Another example of the fact that accuracy is not always paramount is the continuing controversy among journalists about fact-checking stories with sources before the stories are published or broadcast. One proponent of the practice, Boyce Rensberger, formerly of the *Washington Post,* says that such accuracy checks are needed in light of the increasing complexity of science: "If I've got a fact wrong, I definitely want it changed. So I started faxing portions of stories and sometimes entire stories to scientists and asking if I had made any errors." But Rensberger is in a distinct minority. Many journalists—both science journalists and journalists of other stripes—shy away from accuracy checks, out of worry that prepublication review would provide an opening for legal or political challenges to their reporting.[11] In other words, other journalistic imperatives (such as editorial independence from sources) can override the imperative for accuracy.

But if all agree that accuracy is important, there is less agreement on what accuracy is. Science and medical journalists appear to view accuracy as having two facets or dimensions: completeness and absence of factual errors.

Science journalists' concerns about accuracy, and how it is influenced by competitive concerns, are illustrated by their reactions to a controversial story about angiostatin, a potential antitumor drug, published by the *New York Times* by one of its science reporters, Gina Kolata. A former correspondent for the news section of *Science,* Kolata regularly breaks stories that other science journalists must subsequently chase. However, she has been criticized for journalism that is said to be inaccurate and biased. In May 1998, Kolata caused a major stir among science journalists (not to say among the general public): in a front-page story in the paper's Sunday edition, Kolata reported that a cancer therapy that aims at constricting blood flow to tumors, developed by a researcher named Judah Folkman, had shown highly promising results in experiments with rats. The story included a spectacular quote from James D. Watson, a Nobel laureate: "Judah is going to cure cancer in two years." Kolata's story sent other science and medical journalists scrambling, triggering a wave of media attention to Folkman's research, while some journalists criticized Kolata's story for hyping the researcher's work. Watson himself wrote a letter to the *Times* complaining that, although he believes Folkman's work to be promising, Kolata misquoted him from a conversation they had had at a dinner party six weeks earlier.[12]

Many science journalists took exception to Kolata's coverage, complaining about both factual errors and its completeness. Jane Brody, who was Kolata's colleague as a health columnist for the *Times,* criticized Kolata's angiostatin coverage in a speech titled "The Challenges of Accurate Health Reporting." "The competition to get it fast and get it first is fostering a carelessness and haste in health science reporting that had reached an all-time low about a decade ago. Many studies are being brought screamingly to the public's attention long before they should be," Brody said. In the past, Brody said, a story such as Kolata's "would have appeared laden with caveats on a back page of the paper, not as a two-column story on the top of page one." And Brody took Kolata to task for raising readers' hopes that the experiments would lead to medical treatments for humans: "More than two years have passed and still there are no angiostatin drugs curing cancer in people."[13]

Similar reactions to Kolata's reporting were evident in an e-mail discussion list sponsored by the National Association of Science Writers. Two scholars who analyzed the twelve-day discussion found that 83 percent of the critiques posted by participants were negative, with the most frequent criticisms touching on the significance of the story, the story's balance and accuracy, and the placement of the story on the front page.[14]

What Is Accuracy?

Most definitions of accuracy in use by scientists, science journalists, and science-accuracy researchers appear to be rooted in a transmission understanding of mass communication, which holds that a sender uses a medium to transmit a message to a receiver or receivers. This was most famously expressed by Harold Lasswell in his formulation

Who
Says What

In Which Channel
To Whom
With What Effect?[15]

Communication scholars' understanding of the communication process, and therefore of accuracy in communication, was also shaped by the development of information theory in the years following the end of World War II. Claude Elwood Shannon and Warren Weaver, two pioneers in this area, described a communication system as consisting of an information source, a signal, and a receiver. The source provides a message to a transmitter. The message is changed into a signal, which can be distorted through noise, and is received by a receiver, which provides the message to its destination. In Shannon and Weaver's formulation, at its most basic level, accuracy is a measure of the correspondence between the source's message and what arrives at the destination.[16]

However, when news is understood as a social construction, then the notion of accuracy is conceptualized differently. Social constructionism suggests that the news reflects the shared values of those who produce the news—both the journalists and their sources, and, perhaps more indirectly (through the market influence they exercise), readers, viewers, and advertisers. Thus, one theorist of social constructionism has suggested:

> What is this to say of our newspapers, or of eye-witness testimony, or of science? Essentially, when we say that a certain description is "accurate" (as opposed to ... "inaccurate") or "true" (as opposed to "false") we are not judging it according to how well it depicts the world. Rather, we are saying that the words have come to function as "truth telling" within the rules of a particular game—or more generally, according to certain conventions of certain groups.[17]

This approach poses challenges for those who try to assess media coverage of science and medicine. For example, Sharon Dunwoody, a longtime scholar of science communication, has concluded: "I no longer find questions about the 'accuracy' and 'objectivity' of media risk accounts conceptually interesting or worth pursuing; in a world that finally has acknowledged that media accounts are social constructions, answers to the question 'who's right?' are relative."[18]

Thus, the issue of accuracy of news coverage of science and medicine can be understood as an inquiry into the social processes that shape the definition and interpretation of accuracy. That is, the question for consideration should not be: How accurate is news coverage? Rather, the question is: What makes news coverage understood to be accurate, in a given social system?

Take the issue of completeness, which both scientists and science journalists understand to be an element of accurate journalism. News reporting is, by definition, always less than an exhaustive representation of external events, because it is a distilled or compressed version of those events. For example, from all the scientific papers available for coverage at a certain point in time, a science journalist chooses one on which to report; and from that one paper, the journalist culls certain facts for use in his or her media report, leaving unused the bulk of the paper. John Newhagen and Mark Levy have compared the process of generating a media product to the process of making chocolate bars, in which chocolate beans are first concentrated and then used to create many identical confections. "Similarly, in news work, vast amounts of data are gathered from around the world. . . . A similar process of concentration or compression then takes place, where data are reduced to a potent extract or essence called stories, and then combined to create huge numbers of replicas called newspapers."[19] Many science and medical journalists would agree that their work fits this description.

But just as the judgment of whether one chocolate recipe is a subjective one, no one collection of news "essence" is objectively better than any other. Rather, the determination of whether the essence has been correctly prepared—that is, whether the journalists' news stories are sufficiently complete in the face of the data compression that they perform—is an inherently social determination.

Consequently, to the extent that journalists seek to produce accurate journalism, the social processes that define accuracy will heavily influence how journalism is practiced. The issue of accuracy is a contentious one between journalists and scientists, with scientists arguing that mass media accounts are inaccurate and incomplete, while journalists respond that scientists' critiques are impractical for news accounts that are intended for a general audience.[20]

However, the iron grip of the embargo system shows that such arguments, emotional as they sometimes get, are on the margins. In fact, in science and medical journalism, the notion of accuracy has been defined—that is, constructed by scientists, science journalists, and many of the communication scholars who study the interaction between the two—to mean accuracy as determined by scientist sources. "Science is having renewed success in convincing journalists that accuracy is a characteristic of science information that can only be evaluated by science itself," says the science-communication researcher Sharon Dunwoody.[21] Clearly, the embargo system, which exalts the pursuit of factual accuracy over other journalistic values such as enterprise reporting and independence from sources, reinforces this view of how the accuracy of science reporting is to be assessed.

Measuring Accuracy

Mass communication scholars have pursued a number of different ways for assessing the accuracy of news in general and news about science and medicine in particular. Most reflect a transmission, nonconstructivist understanding of communication, in that they evaluate the informational content of media coverage against a supposedly objective standard of accuracy.

In early studies of journalistic accuracy, communication scholars asked the people quoted in news articles about the accuracy of those news stories. In one pioneering study, decades ago, articles from Minneapolis daily newspapers were mailed to sources quoted in the articles along with a checklist of possible errors. The sources reported 0.77 errors per story, with about half of the stories being rated as error free. Errors in names, titles, and "meaning" were the most common errors cited.[22]

A similar approach was taken in one of the earliest examinations of the accuracy of science news, in which two prominent biologists assessed the accuracy of news coverage of biological topics by fifteen daily newspapers in June or November 1921. The researchers identified 3,061 articles, and they concluded: "Gross misstatements of facts were not common. . . . The reliability and high standard of the material collected is one of the definite impressions made by the study."[23]

This approach to evaluating science news accuracy was dominant through the last century. When James Tankard Jr. and Michael Ryan asked scientists to evaluate media stories about their own research, respondents identified an average of 6.2 errors per story. Barbara Moore and Michael Singletary asked scientists to review transcripts of network news stories; more than half rated the stories as less than completely accurate.[24]

But communication scholars came to recognize that scientist sources may operate under biases when asked to identify errors in news stories about themselves. "When accuracy is defined from a source's standpoint, a question always comes up as to whether the criterion for accuracy is 'subjective' or 'objective.'" Indeed, a reanalysis of the data from the survey by Tankard and Ryan found evidence of "objective" and "subjective" dimensions to scientists' evaluations of the accuracy of news media coverage.[25]

Moreover, when Lynn Pulford conducted a news accuracy study that followed the approach used by Tankard and Ryan—with the important exception that Pulford's checklist of possible errors contained 11 items, rather than the 42 that Tankard and Ryan had used—respondents identified 2.2 errors per story, almost one-third the number of errors found by participants in the study by Tankard and Ryan. This finding suggests that the number of errors that a source identifies in a story may be related to the number of errors that the source is asked about, a conclusion that hardly promotes confidence in this approach to error assessment.[26]

Consciously or not, scientists bring their own opinions and experiences to bear when evaluating media accuracy. For example, one scholar recently argued that critics of media coverage of genetics research "applied their own assumptions—usually that any favourable reporting about genetics was undesirable—to condemn all reporting about genetics as bad, simply because much of it contained favourable elements." Robert McCall and Holly Stocking suggested that scientist sources are a poor standard for evaluating the quality of news coverage because they may have been inordinately influenced by bad experiences with the press in general: "It is possible that scientists selectively remember the disasters and forget the successful articles, consider minor errors to be a greater threat to accurate communication than is the case, and are more concerned about how their scientific colleagues perceive them than whether the lay reader has been well-served." In 1982, Sharon Dunwoody pointed out that many inaccuracies identified by scientists in media coverage are not really inaccuracies in the sense of misstated facts, but rather are charges of incompleteness: scientists assess news accounts by the standards of detail used in scholarly journals, and the news accounts are inevitably seen as wanting, she concluded. Dunwoody suggested that errors in science reporting could be classified as either objective inaccuracies, or "mistakes that can be recognized as such by all parties," and subjective inaccuracies, or errors in meaning.[27]

Laura Carsten and Deborah Illman have recently gone further, proposing five types of errors in science reporting: minor corrections, which do not alter meaning; objective technical errors, such as errors in numbers or stating scientific facts; subjective errors, or language that has meaning about which experts might disagree with one another; lack of completeness, such as omitting aspects of a scientific explanation; and style and usage errors.[28]

Faced with mounting evidence that scientists are unreliable standards for evaluating the accuracy of media reports about science and medicine, science communication researchers turned to a different approach. Rather than asking human sources such as scientists to assess the accuracy of news coverage of their research, science communication scholars compared the news coverage to the original published research reports (or other document) on which the news coverage was based. The communication researchers, rather than the sources, identified errors in the news coverage. Much of such research on news accuracy, even in recent years, has dealt with coverage of biomedical topics. Certainly, in large part, this is because of the importance of reporting on risk-related topics. But it also appears to reflect the fact that many health communication researchers are still strongly influenced by a transmission model of communication, in which such an understanding of accuracy makes logical sense.[29]

For example, Eleanor Singer compared media reports of hazard-related research with the original research reports and found that about 40 percent of the media reports differed from the research articles. One analysis of 116 articles in the popular press on breast cancer and mammography found 42 errors in 60 citations to published research. The errors included factual misstatements, treating speculation as fact, and overgeneralizing the research findings. Another study compared the U.S. government's published recommendations regarding mammography with newspaper coverage of the subject over a six-and-a-half-year period. The newspaper articles often took a favorable stance toward mammograms for women between the ages of forty and forty-nine, although this was not a settled issue among experts at the time. Raymond Ankney, Patricia Heilman, and Jacob Kolff found an average of 5.61 errors per daily newspaper article in coverage of a Pennsylvania government study of mortality among patients who had undergone heart bypass surgery at hospitals in the state. Another study evaluated the reporting by five major U.S. newspapers on nutrition-related research published in scholarly journals and concluded that 22 percent of the stories were inaccurate. Other researchers found many errors in 40 Canadian newspaper articles about cancer research. The study identified erroneous information in 55 percent of the articles. In addition, 55 percent omitted qualifications to the research findings, and 18 percent treated speculation as fact. One-fifth of the articles each contained 4 separate errors.[30]

However, not all evidence suggests that news coverage is highly inaccurate when evaluated in this manner. One study—which examined 627 newspaper articles published from 1995 to 2001 in the United States, Canada, Britain, and Australia that collectively covered 111 journal articles—found that only 11 percent of the newspaper articles made moderately or highly exaggerated claims. In addition, 82 percent of the newspaper articles were deemed to have no technical errors. Another researcher, commenting on the project, said: "Their results indicate that media reports are reasonably accurate, except in specific types of controversial areas, and that cases of inaccuracy may be as much a product of the researcher's overenthusiasm as of error by the reporter."[31]

Britain's National Health Service has gone so far as to conduct an ongoing critique of the accuracy of biomedical news coverage in that nation's popular media. The detailed critiques of the accuracy of media coverage of selected biomedical studies are posted online on a Web site named "Hitting the Headlines" (http://www.nelh.nhs.uk/hth/archive.asp).

One group of medical specialists has even developed an "index of scientific quality," or a method for scoring the quality of mass media coverage of biomedical research. The index evaluates the media report on elements such as whether it states the population for whom the research applies; distinguishes facts and opinion; clearly presents the researchers' evidence; reports the magnitude, probability, and consequences of the findings; and assesses the consistency of the researchers' evidence and their conclusions. This index has been used to evaluate medical advice columns for elderly readers, published in Canadian daily newspapers. Half of the columns provided inappropriate information, and 28 percent provided information that was potentially life threatening. Another group, of genetics researchers, developed their own method for assessing the accuracy and balance of stories about genetics research. Scientists and journalists surveyed by the researchers largely agreed on the elements that were essential to be included in the assessment.[32]

In their critique of how the media report about science, David Murray, Joel Schwartz, and Robert Lichter describe a variety of errors that, they say, journalists make with alarming frequency. For example, journalists fail to recognize when researchers with a policy agenda use terms in a way that does not comport with everyday usage—such as making claims about the extent of "spousal abuse" on the basis of data that include minor incidents that many people would not think rises to the severity of "abuse." Journalists, they say, also unquestioningly accept conclusions of research that measures phenomena only indirectly, such as a study about hunger in the United States that asked people whether they had gone without food rather than measuring directly how many people had eaten. In addition, journalists often seize upon one interpretation of a set of data (such as a finding that the number of two-parent families had increased) and ignore another equally valid interpretation (such as a conclusion that traditional families were becoming less common, because the growth in single-parent families outstripped the increase in two-parent families). Indeed, Murray, Schwartz, and Lichter argue that journalists see the world as black-and-white, filled with heroes, villains, and dangers, and that this viewpoint drives them to exaggerate risks, including the risks reported in research. "News that alerts us to dangers, thereby demanding our attention, and that can be shaped so as to conform to the moral world, thereby commanding our affirmation, is a surefire hit."[33]

Communication scholars have also evaluated the completeness of reporting about science and medicine. One study examined science coverage in prestigious newspapers (the *New York Times* and *Philadelphia Inquirer*) and tabloid newspapers (the *National Enquirer* and the *Star*) in September 1987. Although both offered extensive science coverage, that coverage largely focused on findings and ignored the research methods and limitations, which tended to portray science as "a disembodied enterprise." Similarly, an analysis of science coverage by the *Chicago Tribune, New York Times,* and *Washington Post* in the late 1960s, 1970s, and 1980s found that many stories were thin on methodological detail.[34]

When it comes to medical reporting, some scholars argue, missing methodological details can needlessly alarm the public. One former medical reporter notes the public confusion that followed the July 2002 paper in the *Journal of the American Medical Association,* described in Chapter 1, that reported that hormone-replacement therapy did more harm than good to women. That study appeared to contradict many earlier studies that found benefits to hormone-replacement therapy. But there was no contradiction, and if journalists had not omitted a key methodological detail, readers would not have seen one: "Many reporters routinely failed to inform readers that nearly all previous studies were *observational* studies. Observational studies, as a well-educated health journalist knows, are a considerably weaker way to assess a treatment's risks and benefits. That's why the [2002] HRT trial, a well-designed clinical trial with a large number of participants, offered strong enough evidence for doctors and scientists to halt the study in midstream."[35]

Medical researchers frequently lodge such complaints about news coverage. One research project found that U.S. newspapers and network television provided inadequate information on the benefits, risks, and costs of three new pharmaceuticals. A similar project, focused on Canadian newspaper coverage of five new pharmaceuticals, found that coverage always cited benefits of a new drug but usually did not mention possible side effects, and the newspaper articles rarely provided any estimate of the magnitude of the benefits and risks of the pharmaceuticals.[36]

Indeed, the media's handling of the concept of risk is a particularly touchy issue with medical researchers, who argue that incomplete news coverage often exaggerates the risk (or benefits) found in a particular study. Many studies report changes in risk as a percentage: a drug cuts the risk of disease by, say, 50 percent. But reporting the risk in this way is incomplete if the reader is not told how big the risk actually is to start with. "Consider one drug that lowers disease risk from 20 percent to 10 percent, and another drug that lowers it from 0.000002 percent to 0.000001 percent," according to two critics of mass media reporting of medicine. "Both yield a '50 percent reduction,' yet the two drugs differ dramatically in clinical importance." One physician argues that, in just this way, newspapers exaggerated the importance of a study of a new test for a protein that could predict cardiovascular disease, published in the *New England Journal of Medicine* in 2002. Women in the study with higher levels of the protein were more likely to have developed cardiovascular disease over an eight-year period than did women with low levels of the protein. However—and most newspapers didn't report this detail—even the women with high levels of the protein had a low overall risk of cardiovascular disease, so the test would actually detect very few cases of future disease. Such omissions by journalists are the norm. One study of newspaper coverage of three drugs found that only 17 percent of the stories that placed a number on a drug's benefits described the absolute risks or benefits; the other 83 percent described only the relative benefits or risks.[37]

Largely ignored in the foregoing approaches to measuring accuracy is any consideration of how individual readers and viewers process mass media messages about science and medicine. Some factual errors, although unfortunate, may have no negative effect on the public, such as misstating a researcher's name or the journal in which a paper is published. Or perhaps readers and viewers get a correct overall understanding of a science story despite isolated misstated facts in a media report.

Indeed, almost a half century ago, Warren Weaver—the eminent mathematician who helped pioneer a mathematical understanding of the nature of communication and, by extension, accuracy in communication—exhorted scientists against focusing too much on isolated factual errors in news media coverage of their research. Instead, scientists should focus on "communicative accuracy," he said. A passage has communicative accuracy for a given group, he said, if it "take[s] the audience closer to a correct understanding" and contains no misleading inaccuracies that "will block subsequent and further progress toward the truth." Weaver emphasized that these two tests are to be evaluated from the point of view of the audience, not that of experts.[38]

This understanding of accuracy has received much less attention from communication scholars. In one notable exception, mass communication researchers asked individuals to read science stories and then asked scientists quoted in the stories to evaluate the accuracy of the readers' understanding of the articles; more than a third were judged inaccurate. Although scientists were the arbiters of accuracy in this study, they were assessing readers' understanding, not the media reports, and so this study more closely approximated a test of communicative accuracy than the studies reviewed earlier in this chapter.[39]

It seems clear that many of the errors noted by scientist sources and communication scholars who themselves have judged media reports might be seen as harmless from the perspective of communicative accuracy. For example, consider the two errors by the *Washington Post* in reporting the 1999 study on hormone-replacement therapy, cited at the start of this chapter. It is entirely arguable that despite the two indisputable numerical errors, the story remained high in communicative accuracy. Although the article misstated the difference in death rates between women who used hormones and those who didn't, the article still indicated to readers that hormone users had a sharply lower death rate than nonusers. The confusion over the difference in death rates was unfortunate, but it did not mislead readers about the general pattern that the researchers had uncovered in the data.

Also, the completeness of a story may not influence communicative accuracy because readers or viewers may not recall or understand certain details. For example, in one study, college students were presented with newspaper stories based on surveys. Different versions of the stories contained different details about the surveys' methods. Those who read a story with methodological details tended not to remember them afterward, and those who read the details understood the details no better than those who read articles without the details.[40]

Embargoes and Accuracy

However one measures accuracy, does the embargo actually lead to journalism about science and medicine that is more accurate? Certainly, embargo supporters think so. But despite such assertions, little attention has been paid to ascertaining the effects of the embargo on the quality of science and medical journalism. In the study that bears most closely on the question, Fred Molitor analyzed the news coverage produced by five major U.S. newspapers during a 1988 incident in which the embargo system broke down and journalists rushed to report details of a medical study on the use of aspirin to prevent heart attacks before the official embargo release time set by the *New England Journal of Medicine.* His content analysis of news coverage of the study by the five U.S. daily newspapers with the greatest circulation—the *Wall Street Journal, USA Today, New York Daily News, Los Angeles Times,* and *New York Times*—identified errors of omission, inference, and sensationalism in the stories. For example, three of the newspapers omitted the fact that the physicians who participated in the study had been between ages forty and eighty-four, an important fact because the study's results might not hold for individuals of other ages. Indeed, the aspirin study may well have been unrepresentative even of men of that age group, because the physicians in the study smoked cigarettes at less than half the rate of American men in general. But *USA Today*'s coverage of the aspirin study, for example, suggested at various points that the results applied to "most men," "some women," "healthy people," and "most healthy middle-aged men." However, it is important to note, the analysis of news coverage did not identify any errors that were specifically related to the premature release of the story and the ensuing rush to produce news stories.[41]

There were similar complaints about inaccurate news coverage of a 1999 report by the Institute of Medicine on the issue of errors by physicians and other health care workers. Although the report was distributed to reporters under an embargo, NBC obtained a copy of the document through another channel and broadcast a story about the report two days before the scheduled time. Other journalists rushed to cover the report, even though the institute had not yet held a scheduled press conference to explain the report. A television journalist has complained that news accounts misinterpreted the report's estimates of the number of deaths from medical errors, which gave the estimates greater credence than they merited and misreported their meaning. Further, the journalist complains, much of the reporting blamed the medical errors on medical malpractice, even though the report meant the term to include a wide variety of errors. Third, according to this critique, journalists missed the report's emphasis on addressing systemic problems in health care delivery that allowed many of the errors to occur.[42]

In both cases, the fact that one media outlet disseminated its story prior to the embargo release time led other journalists to disseminate their stories quickly. However, in neither case is it clear that the broken embargo was the cause of the journalistic errors. More instructive would be a case in which journalists had no embargoed access to a paper and then had to suddenly rush to produce news coverage of a study that unexpectedly surfaced. This is the scenario that many science journalists worry would develop if journals were to stop offering embargoed materials.

Such a quick-and-dirty situation arose in August 1996, when the aerospace-industry publication *Space News* published an article stating that the National Aeronautics and Space Administration had detected signs of life in a meteorite from Mars. The *Space News* article was published after *Science* had accepted a paper from the researchers but before embargoed copies of the *Science* paper had been distributed to journalists. After the news broke and a media feeding frenzy began, the American Association for the Advancement of Science distributed copies of the paper to journalists with no embargo on the paper, allowing them to report on it immediately. That evening, network news broadcasts carried news of the discovery, as did many newspapers the next morning. Despite the complexity of the topic and the fact that journalists did not have the usual weeklong embargo to research and write their coverage, several key participants in the story found the first-day stories in American newspapers and television to be accurate and to contain appropriate caveats and cautions about the scientists' conclusions. The early coverage was sparse in detail and lacked extensive reactions from other scientists, but these early reports nevertheless were accurate in what they did report: that scientists believed they had found microfossils in a chunk of rock from Mars, and that the finding was controversial among scientists in the field.[43]

Others have reached different conclusions about the accuracy of coverage of the Mars meteorite. One of the researchers involved in the project believes that news coverage misled the public into believing that the conclusions were more sound than they were. "Perhaps the most obvious lesson from this latest chapter in the search for life on Mars is one all too familiar: initially, at least, headlines and soundbites win while facts and reason lose. Most Americans (more than 60 percent by one poll) agree that 'NASA has proved primitive life was present on Mars.'" One journalist has argued that reporters initially overlooked factors that cast considerable doubt on the researchers' conclusions. Journalists would have had the time to become more aware of these contrary facts—and consequently would have been more skeptical in their reporting—if the news had been disseminated under embargo, as planned, Charles Seife said. R. M. Holliman concluded that the first three days of British coverage of the finding were overly simplified and "characterized by an emphasis on 'fossil evidence' at the expense of other evidence which both supports and counters the possibility that microbial life-forms ever existed on Mars."[44]

The meteorite incident also illustrates that assessment of communicative accuracy is inherently subjective and therefore open to debate. On the day when the *Science* paper was released, a CNN correspondent described the putative bacteria in this way: "They looked like something like maggots." The simile may have evoked an accurate visual image, in that the structures do have a rumpled tubular shape, but there also are important differences between maggots and bacteria: the bacteria were microscopic in size while maggots are visible to the naked eye, and maggots are multicellular while the bacteria were not. In this light, does CNN's description meet Weaver's tests of taking the viewer closer to a correct understanding and of not committing errors that would impede further understanding? I believe that CNN's description was communicatively accurate. But Richard Zare, one of the scientists who worked on the project, has a very different view, taking sharp exception to the comparison. It "implies a very, very different situation than what we suggested we might possibly have found."[45]

Further complicating an assessment of the impact of the embargo on journalistic errors is the fact that journalists can and do commit errors even when operating under an embargo. One example occurred in February 1999, when newspaper and television reporters misreported the statistical results of a study, published in the *New England Journal of Medicine,* into the frequency with which physicians referred patients for cardiac catheterization. The paper reported a statistic called an "odds ratio," or the proportion of the odds that blacks would be referred divided by the odds of whites being referred, as being 0.6. The embargo for the paper held, so journalists were able to take the full embargo period for researching and writing their news accounts, but journalists nevertheless erroneously reported that the study had found that blacks were referred for the procedure only 60 percent as frequently as whites. In truth, blacks were referred 93 percent as frequently as whites, and journalists misunderstood and misreported the odds ratio. Interestingly enough, even the paper's own authors misstated the statistical conclusion at one point.[46]

By the measure of published corrections, it appears that errors are relatively uncommon in news coverage of scientific papers that is produced under embargoes. Chapter 1 reported that, during a recent one-year period, twenty-five daily newspapers carried breaking-news coverage of the *Journal of the American Medical Association,* the *New England Journal of Medicine, Nature,* and *Science.* The error rates for those stories were very low, at least according to the newspapers' own standards: the files for those articles in the Lexis-Nexis database show that the newspapers subsequently ran corrections for only fifteen of those articles, including the correction cited at the opening of this chapter. Six of the corrections related to coverage of the *New England Journal of Medicine,* five were for coverage of the *Journal of the American Medical Association,* and two each were for coverage of *Nature* and *Science.* The errors included misleading headlines, errors in names of authors and journals, incorrectly reporting details of a study, and incorrectly reporting the therapeutic implications of a study. However, caution should be used in interpreting this low rate of published corrections as confirming that the embargo promotes accurate journalism. Newspapers are well known to vary widely in their policies regarding the correction of errors; one newspaper might consider a particular passage to be an error that requires correction, whereas another newspaper might conclude that the passage was not erroneous or was erroneous but does not merit correction. In addition, journalists might be unaware of some errors, because sources or readers do not call the errors to their attention.[47]

Proponents of the embargo argue, in part, that reporting about science and medicine is difficult and that the extra time afforded by the embargo gives reporters an opportunity to better understand the story and prepare an accurate account of it. And scientists and journalists certainly have a difficult task.[48] Reading and understanding a journal article can be a challenge on a deadline, because journal articles are difficult to cognitively process. One researcher found that the writing in *Science* and *Nature,* in the 1800s and the first half of the 1900s, was not difficult to read—about the complexity of an English-language daily newspaper. But after 1947 for *Nature* and after 1960 for *Science,* the two journals became far more difficult to read. Ten other scientific journals also became more difficult to comprehend between 1900 (or the year that the journal was founded) and 1990, by which time all ten journals were far harder to understand than a daily newspaper.[49]

It is not just reading the text of papers that is difficult. Much modern science is expressed through statistics, and many journalists have difficulty with mathematics. In one recent survey of health reporters at midwestern newspapers, about a third of respondents said they usually had difficulty in interpreting statistical data. The same proportion reported difficulty in understanding health issues. (By contrast, the respondents reported less difficulty in finding sources, putting health news into context, and producing balanced stories on deadline.)[50]

Moreover, there is extensive evidence that humans perform difficult cognitive tasks less well when they are under stress, such as the time pressure experienced by a journalist under deadline. In 1908, two researchers tested the ability of mice to detect the difference in brightness in two boxes, under varying levels of stress caused by electrical shocks. The researchers found an inverted-U relationship: increasing stress levels caused better learning in the mice, but only to a point, after which increasing the stress even more led to decreases in learning. This relationship is sometimes called the Yerkes-Dodson Law after the two researchers who first described the relationship, and it has come to be held to explain human performance under types of arousal including stress.

The commonly accepted explanation for the inverted-U shape of the relationship is that stress narrows the subject's attention, much like a spotlight narrowing the field that it illuminates. Some narrowing of attention is helpful, but too much is harmful. In reaching a decision, for example, moderate amounts of stress would improve cognitive performance, because the narrowing process at first leads the subject to exclude extraneous information. But higher amounts of stress would hurt performance because the additional attention would lead the subject to start excluding pertinent information in making the decision.[51] Applied to the embargo system, this line of thinking might seem to suggest that the absence of an embargo would hurt the accuracy of science journalism by increasing deadline pressure on journalists. Journalists certainly perceive that deadlines do increase the stress under which they operate.[52]

However, the validity of the Yerkes-Dodson Law has come under increasing question by cognitive scientists, who find it theoretically and methodologically lacking.[53] Moreover, there is evidence that, in experts, the attention-narrowing process tends to focus attention on decision-making techniques that the expert would habitually use and exclude other techniques. As two scholars concluded in one review of the implications of cognitive science for mass communication research: "It might be that journalistic ways of doing things are so hardwired that reporters, no matter how much time they have, operate much the same way, purely out of habit."[54]

Indeed, the notion that the embargo will improve journalistic accuracy is premised on the assumption that when offered more time to prepare their coverage of science and medicine, journalists will use that additional time. Concomitantly, embargo proponents believe, the elimination of embargoes would create a weekly stampede by science journalists to quickly report on important journal articles—a stampede in which accuracy could suffer. But the validity of both these assumptions is unclear, at best.

Certainly, some journalists say that, in some or most situations, they do take advantage of the extra time offered them under the embargo. But an individual reporter may not use all the additional time that the embargo provides. With an embargo of several days, the reporter may work on the embargoed story in bits and pieces, fitting that story around other stories that the reporter is covering. (As an aside, I would note that this observation runs counter to an occasional claim that science journalists embrace embargoes because they are lazy. Journalists invest a significant amount of time in identifying, researching, and writing news reports about embargoed journal articles; laziness does not appear to me to be a motivating factor in the popularity of embargoes. The journalists appear more interested in seeking ways to reduce uncertainty in their selection of stories than in reducing their workloads per se. To my mind—as is explored in greater depth in this book's last chapter—one of the major problems with the embargo is that it misdirects journalists' efforts, not that it capitalizes on laziness.)

In some cases, journalists use the embargo as a time-management device for planning their week and advising their editors about major science and medical stories that they may see in the competition. Moreover, although scientists seem to equate "the media" with the reporters who write or broadcast stories about science, media organizations actually are large organizations, consisting of many individuals with roles in the preparation and dissemination of news reports about science and medicine. Many parts of the media-production process operate the same regardless of whether an article or broadcast report is based on embargoed information, and therefore the embargo cannot influence the accuracy of these aspects of the production process. For example, in the case of breaking-news coverage of embargoed journals by newspapers, a news article is likely to be edited during the afternoon or evening before it runs in print, regardless of how many days the reporter spent researching and writing the article. That article is edited by (perhaps several) editors, and each may have the authority to alter the article without the author's review or consent. An editor, rather than the reporter, will decide on which page the article will be placed, and may cut material from the article for it to fit the allotted space, again without consulting the reporter. Headlines and photo captions are likewise written by editors, not the reporter, on the afternoon or evening before the newspaper is published. As one physician-turned-reporter-turned–journal editor has written: "Editors sometimes shorten reports about medicine, and important information may be omitted. A reader has no way of knowing whether the reporter failed to include the relevant information or an editor cut it out."[55] This underscores the fact that not all errors can be laid at the foot of the reporter whose byline appears on the article, and therefore these errors are not influenced by the extended period of researching and writing offered by the embargo.

Conclusions

The point of this chapter has not been to endorse errors in journalism. Particularly in medical journalism, erroneous reporting can be life threatening, if it leads readers or viewers to reach medical decisions, such as halting the use of a medication, without consulting with a physician. Even in reporting on nonmedical topics, journalists have an ethical duty to be accurate. This chapter has sought to emphasize that judgments of accuracy are inherently subjective and social in nature. Proponents of the embargo system maintain that embargoes promote journalistic accuracy, but this claim is essentially tautological, because the embargo system reflects and fosters a definition of accuracy promoted by the scientific establishment. However, even if it is conceded that the embargo makes journalism more accurate, one must weigh the benefits of this added accuracy against the embargo's costs to society. Those costs will be considered in the next chapter.

4

Costs of the Embargo

The preceding chapters have examined the workings of science journalism, how journal embargoes influence the research papers that media organizations cover as news, and how journal embargoes influence journalists' understanding of accuracy in their work. This chapter will examine several ways in which the embargo has unintended—and undesirable—effects on journalists, the public, and society as a whole.

Impact on Journalists' Coverage of Science

One major problem with the embargo system is the way it distorts journalistic reporting about science to highlight the "latest" findings, regardless of their true significance. The embargo system creates artificial urgency regarding the papers published in a handful of journals; many other journals that may contain research papers of comparable merit and importance go untapped by journalists simply because they do not make journalists' lives easier by providing embargoed copies.

Some critics of the embargo system have caricatured science journalists who participate in it as lazy. In fact, science and medical journalists invest an enormous amount of time and effort each week in trolling through the embargoed materials, in search of papers that they should cover—because of the papers' intrinsic news value, because other news media are likely to cover the papers, or (most likely) because of some combination of both reasons. But journalists have only a finite amount of time and energy, and devoting so much of it to covering embargoed journals inevitably prevents them from covering other aspects of science and medicine.

Indeed, the scientific establishment believes that it benefits from the way in which the embargo channels science and medical journalists into covering the latest developments. As the historical account in Chapter 3 documents, scientific societies in particular have come to embrace the embargo as having value for science as well as for the public. Floyd E. Bloom, a neuroscientist who served as editor in chief of *Science* from 1995 to 2000, says that science as a whole benefits from the increased press attention created by the embargo: "Wide coverage of scientific discoveries is good for the public and the scientific community because it illustrates the intellectual and practical value of publicly funded investment in scientific research."[1]

But by influencing how journalists use their time, embargoes on scientific journals also create winners and losers among scientific disciplines and scientific associations. The winners are the disciplines that secure a high amount of press coverage through the embargo, such as medicine. The losers are disciplines that do not impose an embargo on their scholarly journals and therefore lack the cachet of simulated newsworthiness.

And there are many disciplines that do not embargo their journals. Among the scholarly societies in this category are the American Geophysical Union, which is devoted to earth science and astronomy; the American Physical Society, an organization of physicists; the Association for Computing Machinery, a computer science group; and the Institute of Electrical and Electronics Engineers, an electrical engineering society. These scientific groups pay a penalty for eschewing embargoes: their journals get much less press coverage than the embargoed journals do.

For example, the American Geophysical Union annually publishes more than fifty thousand pages of peer-reviewed scientific research in its journals, including the *Journal of Geophysical Research* and *Geophysical Research Letters.* Like the publishers of the embargoed journals, the AGU sends press releases to journalists to highlight the most newsworthy of those papers. But unlike the publishers of embargoed journals, the geophysical union does not follow a regular, weekly routine for its releases. The geophysical group, maintaining that its goal should be to disseminate scientific findings as widely and as quickly as possible, also does not embargo the press releases or the journal articles upon which they are based; journalists are free to report on them immediately.

One might think that journalists would jump at the chance. To the contrary, the geophysical union's journals receive less press coverage than do the elite embargoed journals, says the society's public information director, Harvey Leifert. He argues that *Science* and *Nature* have managed to essentially guarantee themselves a certain amount of newspaper space and television coverage every week. Because of the embargo's predictable schedule, he says, editors and producers know that their competitors are likely to be covering those same journals on certain days, and they therefore want to do the same. By contrast, the nonembargoed journals have no such guarantee. "Stories based on AGU journals have to fight for space, not only with other science news, but with other news in general. We do envy the automatic access *Science* and *Nature* secure via their embargo process. If there were no embargoes, editors would have to use news judgment to determine daily which science stories to carry."[2]

Science journalists would object—correctly—that embargoed journals do not really have places reserved for them in newspapers and television broadcasts. For one thing, multidisciplinary journals like *Science* and *Nature* publish articles from a broad spectrum of research fields and so, in any week, may be more likely to offer newsworthy research than do single-discipline journals such as the geophysical union's. Moreover, television coverage even of the embargoed journals is spotty at best. During the fifty-two weeks from June 1997 to May 1998, for example, *Science* received coverage only four weeks by any of the U.S. broadcast networks' evening news broadcasts; *Nature* was covered in only three weeks. During the same time span, the AGU's *Journal of Geophysical Research* and *Geophysical Research Letters* received no coverage at all by the network evening broadcasts. (The embargoed medical journals that are best known in the United States—the *Journal of the American Medical Association* and the *New England Journal of Medicine*—fared better, scoring breaking-news broadcast coverage in eighteen and twenty-three weeks, respectively. But these medical journals, because their disciplines are so far removed from earth sciences, are not likely to directly compete with the AGU's journals for coverage.)

But with respect to newspapers, Leifert seems closer to the mark. In the same fifty-two-week period, there were only two weeks in which *Science* drew breaking-news coverage from none of the twenty-five daily newspapers in the study, and three weeks in which *Nature* received no breaking-news coverage. (The *New England Journal of Medicine* and the *Journal of the American Medical Association* had newspaper coverage of each one of their issues.) By contrast, the AGU's two journals were covered in only ten articles published in any of the twenty-five newspapers during the same period.[3] In short, embargoed science appears to have a clear edge over unembargoed science, in getting into the paper and on the air, regardless of the relative importance of the two.

Although they never say so, some leaders of the scientific establishment may be equally happy that the embargo, and the demands that it places on journalists to devote time and effort to covering the "latest" research, steers most journalists away from an investigative or critical approach to covering the scientific establishment. Such endeavors would reduce a journalist's available time for covering the embargoed journals. "The often slavish reliance on a few journals implies taking science as a given, simply reporting on work that is already done," says Jon Turney, a former science journalist who now is a science communication scholar. "With a supply of easy stories guaranteed, there is little incentive to ask about issues like the motivations underlying funding or who creates the agenda for doing the research."[4]

There are international aspects to this phenomenon as well. The embargo system, which is dominated by journals published in the United States and Britain, draws the attention of science journalists in other countries away from the journals published in their own nations. In one study, Dutch medical journalists said that they did not monitor journals published in France or Germany, because they believed that any important research would appear in a journal published in the United States or Britain.[5] "This Anglo-American supremacy conditions the very definition of science news, and its content," one scholar has observed.[6] Indeed, a content analysis of fifty years of science coverage in Italian newspapers found "almost no reference" to articles published in Italian journals but numerous articles based on journals published in other nations.[7] As noted in Chapter 1, European research agencies started their own Web-based system for distributing embargoed and nonembargoed materials to journalists, as a counterweight to EurekAlert! But the problem continues for journalists in other areas of the world. One Brazilian science journalist, for example, notes that he and his colleagues rely heavily on EurekAlert! and *Nature*'s press site and that

> there are too few stories about Brazilian scientific research in Brazilian newspapers and magazines. . . . To gather such material, reporters would have to keep tabs on dozens of scattered labs, universities and research institutes, and this is not practical to do given our staffing. For us, it is easier to learn about a newsworthy piece of research done by a local scientist when the findings are published in, for example, . . . *Proceedings of the National Academy of Sciences.*[8]

In short, Brazilian journalists construct research as news only after a major journal outside that nation has validated the science; other research projects remain nonnewsworthy and therefore unreported. One might suggest that it is not the business of American science journalists to worry about the impact of their reporting methods on science journalists in other countries. But even so, scientific institutions that ostensibly want to promote an understanding of science in all nations and cultures should give close consideration to the transborder impact of news embargoes.

Overall, one of the most troubling aspects of the embargo system is the fashion in which it discourages journalists from competing with one another. Miriam Shuchman and Michael Wilkes contend that the medical journals' embargo promotes a "limited focus" by reporters on embargoed journals. "Medical journals promote media overemphasis on single articles through the embargo system," they argue. "The embargo system ... increases the perceived newsworthiness of a journal article, thereby encouraging an over-reliance on journals as a source of news." Shuchman and Wilkes suggest that in the absence of an embargo, journalists might scan a wider range of medical journals or might undertake to follow developments in specific research areas over time.[9]

Such strong influences on journalistic behavior fly in the face of the "marketplace of ideas" view of journalistic activity, such as that advanced by the Hutchins Commission, which held that the normative requirement for freedom of the press is rooted in the competitive nature of the press. Even a former editor of the *New England Journal of Medicine* agrees. The embargo "removes some of the gumption of science writers to go after stories that are more detailed and more in depth and more in context," said Jerome P. Kassirer. "It's too easy to report on the latest study, rather than going into a single subject in detail."[10]

Most journal editors and proponents of the embargo system tend not to directly acknowledge the competition-inhibiting aspect of embargoes. Rather, when touching on this aspect, they speak of the embargo as a means of treating all journalists equally. However, although no one would argue that journals should play favorites with journalists by releasing information to some before others get access, the embargo goes beyond ensuring this equity. In addition to orchestrating the dissemination of journal articles to journalists in an organized and predictable way, the embargo also forces all journalists to delay their coverage for as much as a week. In effect, the embargo forces the best science journalists, who might be able to turn around a story on a research paper in an afternoon, to cool their heels to give slower journalists more time to produce their stories. The embargo insulates journalists from the ways in which competition among journalists promotes quality. In the absence of an embargo, more skilled journalists would be able to get their coverage into print, on the air, or online more quickly than journalists with less skill. Or less skilled journalists might publish their coverage as quickly, but it could be incomplete or riddled with errors. The embargo prevents this differentiation among journalists and journalistic organizations from surfacing in public view.

Impact on the Public

The embargo's effects on journalists might be of little more than academic interest, save for the fact that anything that influences journalists may have important ripple effects on the public that watches the journalists' television broadcasts and reads their newspaper reports. In that sense, journalists aren't the only ones affected by the sense of immediacy and importance that the embargo system lends to stories about science and medicine. Readers and listeners are also subjected to the constant barrage, week after week, of news stories touting the latest scientific research.

When journalists concentrate their attention on a few major journals, that can also give the public a distorted picture of current research, because each journal has its own agenda of the types of scientific research that it is willing to publish. By limiting their news coverage to a few journals, journalists effectively limit themselves—and their readers and viewers—to the agendas of those journals. Journalists maintain that they focus their energy on a handful of major journals because these journals publish the most important cutting-edge research. There is, of course, some truth to this: journals such as *Nature, Science,* the *Journal of the American Medical Association,* and the *New England Journal of Medicine* do frequently publish important research, and it is entirely legitimate for journalists to monitor these publications.

But important research is also published in nonembargoed journals, and science and medical journalists may miss these papers if they are wholly focused on the embargoed journals. The Institute for Scientific Information, which compiles information on the papers that scholars cite in their work, has prepared a list of the ninety most commonly cited journals in a total of nine disciplines. Although the embargoed journals headed the list, the institute found dozens of unembargoed journals that exert major influence in their fields. Another study found that, from 1983 to 1993, journalists overlooked a number of important articles, published in nonembargoed journals, on the relationship between alcoholism and genetics. Yet another example: Starting in 1999, a debate raged in the pages of unembargoed physics journals over the possibility of releasing energy from an isotope of the element hafnium, a process that could be the basis of a controversial new weapon of immense power. Journalists largely ignored the debate, with a few notable exceptions.[11]

When journalists draw their news stories from the small pool of elite journals and ignore other journals, they are hitching their news coverage of peer-reviewed research to the publishing agendas of the elite journals. And there are important differences between the elite journals and the journals that reporters rarely cover. For example, one study that compared the articles published in the top embargoed medical journals with the articles published in journals that specialize in women's health issues found that the elite journals had a "narrow definition of women's health." Specifically, most of the women's health articles in the elite journals dealt with female reproductive conditions, such as menopause and female-specific cancers such as ovarian cancer. By contrast, the specialty journals published more articles on topics not tied specifically to reproduction, such as eating disorders and osteoporosis.[12]

This might seem to be a minor problem at worst for news coverage of basic science. How troubling is it, really, if readers get fewer chances to read about high-energy physics or organic chemistry? But this skewing of how journalism covers science is important, because even an accurate rendering of individual research projects can confuse or mislead the public, if the individual reports add up to a confusing or misleading picture of science. Science as an institution depends heavily on the public—through means such as the use of taxes to operate government scientific agencies and universities and to provide grants to researchers. The way in which the news media cover science may help mold public understanding of science.

Scientists do understand this. For example, Floyd Bloom, formerly of *Science,* says that the embargo is good because it promotes media coverage of science, which benefits science: "Media coverage is good for authors because it makes clear to the public, both in funding circles and in the local press, their most recent scientific achievements."[13]

However, even granting that science benefits from media coverage of research because that media coverage promotes public understanding of and support for science, it is not a foregone conclusion that shaping that media coverage through an embargo produces those results. Indeed, the embargo encourages a type of science journalism that depicts research as little more than a series of isolated discoveries, with little connection to previous research and divorced from a systematic mode of investigation. According to Peter Gorner, a science journalist at the *Chicago Tribune:* "The public is probably confused by even the pronouncements from the leading journals because studies are seldom put in context. One week, coffee is said to cause pancreatic cancer, the next week it is said that it doesn't. Or a study tells you what you already knew: A piece in the *Archives of Dermatology* once proved that if you don't wear sunscreen on the beach, you'll get sunburn."[14]

What seems to vex scientific leaders most about public misunderstanding of science is that many members of the public do not understand science as a methodical means of inquiry, with rigorous standards of proof and with one line of inquiry building on another and then another on it. When members of the public don't understand that essence of science—these scientific leaders believe—the public becomes susceptible to pseudoscience. The embargo, by promoting an unending stream of coverage of the "latest" research findings, diverts journalists from covering the process of science, with all its controversies and murkiness.

The fact that news coverage of science does little to bolster individuals' understanding of the process of scientific inquiry is illustrated by analysis of data from a 1999 survey commissioned by the National Science Foundation, in which respondents were asked about their use of news media such as newspapers and were asked various questions to test their knowledge about scientific facts and the method of scientific inquiry. When variables such as gender and education were taken into account, the relationships between media use and knowledge about science were only weakly positive. Newspaper use, for example, had an association of only 0.06 with procedural science knowledge; use of science magazines had an association of 0.09 with procedural knowledge. However, general television use was negatively associated with both factual knowledge about science (−0.10) and procedural knowledge about science (−0.15).[15] Although public understanding of science is a complex phenomenon, undoubtedly influenced by many factors beyond embargoes on mass media reporting on journals, the embargo could be shaping the public's view of science in ways that scientists and journalists would not have sought.

Of course, medical reporting offers additional concerns, because the stories covered by the news media about medical research can affect what members of the public do, whether they decide to kick unhealthy habits like smoking or unsafe sex or demand new medical treatments that they read about in the newspaper. Particularly when it comes to medical research, news coverage of a research finding can prompt patients to demand treatments from their physicians that may be inappropriate or may trigger unwarranted angst or hope for certain patients. In addition, when highly publicized medical studies appear to differ from one another, the publicity can produce confusion and frustration among members of the public who are trying to figure out how to avoid disease or recover from it. Indeed, nutrition researchers are becoming concerned that mass media reporting about medical research is producing an impression among members of the public that medical researchers are inconsistent in their guidance about what constitutes a healthy diet. These researchers worry that the media-inspired confusion will trigger an anything-goes "backlash" among members of the public, with an increase in unhealthy eating habits.[16]

The problem is heightened by the fact that journalists are far more likely to report on a "positive" research finding—such as the conclusion that a drug has an effect in patients—than in a negative finding of no effect. This was illustrated by newspaper coverage of two articles published back-to-back in a 1991 issue of the *Journal of the American Medical Association.* One of the journal articles found that men working at a nuclear facility had an elevated risk of leukemia; the other failed to find an increased leukemia risk among people living near nuclear facilities. The first study was covered by more newspapers than the second, and in stories that described both articles, the one that found the elevated risk got more space. Similarly, over a fifteen-year period, newspapers gave more coverage to research supporting a genetic link to alcoholism than to studies that found evidence questioning such a link.[17]

Marcia Angell, former executive editor of the *New England Journal of Medicine,* says the media's emphasis on reporting only on newsworthy research produces a constant stream of news reports about "startling" research findings. That is a problem for the public, she believes, because "startling research is more likely to be wrong than confirmatory research. Solid conclusions are reached bit by bit. The more studies done of a particular question, the more accurate they are likely to be. . . . News of the entire sequence is unlikely to make it to the media, at least not toward its end, when it is most reliable."[18]

Angell and Jerome Kassirer, while they were editors of the *New England Journal of Medicine,* similarly argued that the media place too much emphasis on individual articles in medical journals: "What medical journals publish is not received wisdom but rather working papers. Each of these is meant to communicate to other researchers and to doctors the results of one study. Each study becomes a piece of a puzzle that, when assembled, will help either to confirm or to refute a hypothesis. Although a study may add to the evidence about a connection between diet or exercise and health, rarely can a single study stand alone as definitive proof."[19] But although Angell and Kassirer contended that the media and public both should be more sophisticated in their consumption of medical research, the two editors ignored the complicity of their journal's embargo policy in fostering this situation.

Steven Rosenberg, chief of surgery at the National Cancer Institute, suggests that the conventions of journalism distort biomedical news. "Medical progress occurs very slowly, over many years, and the need to make it current seems to lead to the creation not of real events but more of pseudo-events that surround things that had nothing to do with the progress itself—a report in a medical journal or a press conference or a scientific meeting."[20]

Indeed, Shannon Brownlee, a veteran science journalist, argues that media coverage of medical research continually emphasizes the hopes that new research may offer for patients, while downplaying any uncertainties. That emphasis on hype and the upbeat, she says, arises from the media's dependence on journal articles as news sources.

> We have this symbiotic relationship with the industry we are supposed to scrutinize, so much so that we often get our story spoon-fed and pre-digested from the medical journals, which send out embargoed copies of the top scientific papers each week. That means the editors and publicists of those journals, rather than reporters, are deciding what constitutes news. Funny, but those digests never seem to include the (rare) editorials that criticize the medical industry.[21]

Impact on Science and Medicine

Members of the general public aren't the only ones who read newspapers and watch television. Although scientific journals are designed for scientists and physicians, these specialists often actually learn about new research through the mass media, long before the journal article describing the research arrives in the mail. Thus, the distorted picture of science that can be blamed on the embargo may also skew the understanding of scientific developments by scientists and physicians.

Although the public learns of information published in scholarly journals (usually through the news media), these journals are actually intended to be read by scientists and physicians. Scientists read journals in their field as part of their use of the scientific method, which builds upon the work of other scholars. A researcher may rely on others' work for the theoretical basis of a research project, for data to build upon, or even for counterpoints that can be challenged or overturned. Physicians monitor medical journals for information on topics such as new therapies, new uses for old therapies, and evidence that older therapies do not work as thought or may have undesirable side effects.

But how does a physician or scholar become aware of work published by other scholars in journals? In other words, how do physicians and scientists learn about new scientific research in their fields? Communication scholars see such a question as one of "news diffusion," or of the processes by which a certain bit of news flows through the community. In his landmark study of diffusion of innovation, Everett Rogers proposed a five-stage process to describe an individual's decision about whether to adopt or reject an innovation. In the first stage, called the knowledge stage, "an individual . . . is exposed to an innovation's existence and gains an understanding of how it functions."[22] It is in this stage that mass media channels, such as television and newspapers, are potentially influential because they can effectively spread information to a wide audience. In subsequent stages, the individual is persuaded to make a decision regarding the innovation, reaches and implements that decision, and then evaluates it.

As Chapter 1 described, scholars use a wide array of information sources to monitor developments in their fields, including informal exchanges with colleagues, scientific conferences, and peer-reviewed scholarly journals. But scholars also rely on the popular news media, particularly for news about research outside their specific fields, which they would be likely to learn about through scholarly channels. A highly publicized example occurred in 1989, when researchers B. Stanley Pons and Martin Fleischmann of the University of Utah, using inexpensive laboratory equipment, produced what they concluded was nuclear fusion at room temperature. If true, this was a dramatic development, contradicting physicists who had said that controlled fusion on earth would require massive apparatuses and temperatures of millions of degrees—and governments around the world had sunk billions of dollars into research projects based on the theories of mainline physicists.

In what he later said was an inadvertent violation of the embargo, Fleischmann disclosed the findings to a journalist (as discussed in Chapter 2), and Pons and Fleischmann discussed their work at a press conference before even submitting a paper with details of their experiments to *Nature*. Scientists around the world, who learned of the cold-fusion results through the mass media, leaped into action, trying to repeat the experiment in order to demonstrate the validity of the claim by Pons and Fleischmann. Traditional means of scientific communication—such as journal articles and presentations at conferences—soon began to play an important role in disseminating research results to cold-fusion scientists and their critics, but the mass media remained an important source in the fast-moving field as well.[23]

This example, though more dramatic than most, demonstrates that scientists do learn about research from television and newspapers. The principle applies for more mundane research as well: When asked in a 1991 survey about nonspecialist sources of information that are important for their work, 57 percent of participating Dutch biologists listed national newspapers, as did 45 percent of Dutch engineers. Thirty percent of the biologists said they relied on Dutch television, as did 20 percent of the engineers.[24]

Physicians, too, use the news media to keep up with developments in medical research. Indeed, one of the rationales offered by scholarly publishers for their efforts to control the timing of news stories about scientific and medical research is to try to guarantee that physicians who learn about medical research in the mass media can immediately read the journal article that was covered by the news media. (However, it is open to question whether physicians really do read medical journals in time to be prepared for media-prompted questions from their patients; for example, female physicians in a recent survey reported spending only about one hour each week reading medical journals.)[25]

It is difficult to assess how a particular piece of research influences the progress of science and medicine. The most common approach is to examine the frequency and pattern of citations of a specific article in articles written by other scholars. And there appears to be a link between news coverage and citation rates of journal articles: twenty-five articles published by the *New England Journal of Medicine* that had been covered by the *New York Times* received more frequent citations by other scholars than thirty-three other articles, also published in the *New England Journal of Medicine,* that the *Times* did not cover. News coverage appeared to have its strongest effect on citation rates in the first year after an article's publication, but there continued to be an association between news coverage and citation rates for at least the subsequent nine years. The researchers extended the analysis to include a twelve-week strike at the *Times* in 1978, when the newspaper was published but not circulated. Journal articles that were covered during this period—when researchers were unable to read the *Times*—were cited no more frequently than other journal articles. David Phillips et al. concluded that the *Times* coverage per se was boosting the prestige of the scientific findings.[26]

Although the Phillips study did not address embargoes, it nevertheless has helped generate support among scientists for the embargo system, because it is very important to scholars for their papers to be cited frequently by others, and therefore they want journals and their institutions to use the embargo to promote coverage by the *New York Times*.[27] One reason for scholars' desire for their work to be cited, of course, is that citations signify that other scholars judge a paper to be an important contribution to scholarly knowledge; generally, it is thought, scholars would not bother to cite a weak or meaningless paper (except for the occasional attempt to debunk or attack it). Scholars are often evaluated on the "impact factor" of the journals in which they have published. A journal's impact factor for a given year—a measure of how frequently its articles are cited by scholars—is computed by dividing the number of times its articles were cited in the previous two years by the total number of papers that the journal published during those two years. High impact factors are often thought to signify that a journal is of high quality, though there are many objections to such a conclusion.[28]

Having a high citation rate, or publishing in a high-impact journal, is important to scholars for many reasons, because citation rates can have all sorts of ripple effects on a scholar's academic career. A scholar who has published highly cited papers is more likely to win a new position, tenure, or promotion than a scholar who has published papers that are cited less frequently. One molecular biologist recently noted: "Scientists are increasingly desperate to publish in a few top journals. . . . If we publish in a top journal we have arrived, if we don't we haven't." Similarly, a researcher with high citation rates is more likely to win grant money; the researcher is seen as having produced high-quality research in the past and so is a good bet for producing more high-quality research with the grant.[29] Colleges sometimes allocate funds among academic departments—or even decide which ones to close and which to keep open—based on the citation rates of the departments' members. Citation rates even are a factor in the rankings of colleges' doctoral programs that are published by the National Academy of Sciences; colleges want to rank highly so that they will attract the best graduate students.

Journal publishers too want their publications to have high impact factors. High impact factors set up a reinforcing cycle, of sorts, between the journal and the scholars who submit papers to it: a journal with a high impact factor is likely to attract the best papers, because scholars hungry for tenure and promotion will want the best chance for their own papers to be cited frequently. One scholar notes: "Editors compete to publish those articles that they think are important or 'hot,' articles that will raise the 'impact factor' of their journals."[30] Those submissions in turn keep the journal's impact factor high. Moreover, college libraries, which have been forced by tight budgets and skyrocketing subscription costs to cut back on journal subscriptions, often take impact factors into account when choosing which subscriptions to cancel.

Although one implication of the Phillips study—for both scientists and journal publishers—is that one way to boost a journal's citation rate is to promote mass media coverage of it, the study examined only the impact of coverage by the *New York Times* and did not address the possibility that coverage by other media might also inform scientists about new research and influence their subsequent reliance on such research. Other news media are likely to have an impact on citation rates as well. Indeed, I have elsewhere examined the citation rates for papers in the principal four embargoed journals—the *Journal of the American Medical Association, Nature,* the *New England Journal of Medicine,* and *Science*—and news coverage of those papers by daily newspapers across the country and the three broadcast networks, during a one-year period. The study found that papers that had been covered by television or newspapers were cited an average of 116.46 times during the following five years, while journal articles that had received no news coverage were cited a mean of 90.52 times. The study also found no association between citation rates and coverage by the *New York Times* or by television evening news broadcasts; the only significant association was with coverage by twenty-four other U.S. daily newspapers, with every one hundred words of newspaper coverage associated with, on average, 1 additional citation of the journal article. This finding suggests that the volume of news coverage in general, rather than the amount of news coverage by elite media outlets such as the *Times,* is a key influence on disseminating news about research to scientists and physicians.[31]

That the mass media have such an important role in informing scholars about scientific research has important policy and research implications regarding the transmission of scientific findings through the popular media. By informing researchers about new research, the media may also unwittingly influence the conduct of science. Laurel Richardson Walum, a sociologist, raised such concerns in 1975 after the *New York Times* ran an article about her research and she received a torrent of letters from colleagues. "To the extent that the press coverage *itself* prejudices the sociological jury (either positively or negatively) regarding a paper, and to the extent that they allow the norms of journalism to decide for them the value of a work, the norms of science are endangered."[32]

Scientific journals could themselves be damaged by their pursuit of publicity. Although one aim of the embargo system is to burnish journals' reputations with scientists, Richard Horton of the *Lancet* argues that the publicity practices of scholarly journals could have the opposite effect, if journal editors decided which papers to publish based on their publicity value rather than their scientific merit. "Editors already allow articles to be selected and précised into suitable word-bites for their press release. To summarize a complex paper that requires careful interpretation and verification—but which is an excellent story journalistically—immediately puts at risk a journal's credibility, since the editor is now making further editorial decisions about quality beyond that of peer review."[33]

This may well be happening. One study documented that seven elite medical journals exaggerated the importance of published research in the press releases that they distributed to reporters. For example, the researchers found that only 23 percent of the press releases described the limitations of the study in question, and only 22 percent of the press releases reported any financial ties between the research and industry.[34] To the degree that reporters are taken in by these exaggerations and omissions and repeat them in their news coverage, physicians and researchers may be getting a distorted sense of the significance of the research reported by the mass media.

Impact on Financial Markets

Although much of the research published by scholarly journals is too basic to influence the stock prices of companies, some research papers do have that potential.[35] For example, a medical journal could report on research into the effectiveness of an experimental drug—findings that could influence whether the federal government would approve use of the drug in humans. Or a biotechnology company may have tested a technology that could serve as the basis of new products. In such cases, the enormous press coverage of such findings, stimulated by the embargo, could be expected to trigger movement in the company's stock price after the embargo lifts and the news coverage is disseminated. Anyone with advance knowledge of the research findings could take steps to profit from those movements in stock prices. Someone with advance knowledge of favorable results of the test of an experimental drug could buy shares of the company's stock, in expectation that the news coverage of the results would boost the stock's price.

The embargo system creates just such an underground economy of secret advance information about commercially important research. Stock analysts and institutional investors (and, at least hypothetically, journalists) can benefit from this system, because they receive the information or are able to ferret it out and are able to act on the information. Smaller investors, such as members of the general public, do not have access to this information and therefore are placed at a comparative disadvantage.

This underground economy arises from two aspects of the embargo system. One is the fact that journalists are given early access to the research—including copies of the texts of the papers—before the public gets that access. The second is that the embargo establishes a well-known time before which there will be no news coverage of the research; anyone who can manage (licitly or illicitly) to learn the details of research before this embargo release time could take steps to profit on that knowledge.

The fact that science and medical journalists are granted confidential advance access to the contents of scholarly journals means that those journalists—and others that they communicate the information to—may violate federal insider-trading laws if they use that information for their own profit. At first glance, such a legal conclusion might seem debatable, since journalists do not work for the companies whose stock is influenced by the research. But the Supreme Court has held that insider-trading laws can indeed be applied against "outsiders" such as journalists. In 1997, the Court ruled that a lawyer violated the insider-trading laws when he made personal stock transactions—netting $4.3 million in profit—based on his knowledge that one of his firm's clients was about to try to take over Pillsbury Company. Even though the lawyer worked for Pillsbury's attempted suitor and not Pillsbury itself, the Court ruled that the insider-trading laws could be applied to an "outsider" who, like the lawyer, misappropriates confidential information to which the outsider has access.[36]

Based on such legal reasoning, it seems clear that journalists would run afoul of the law by making investment decisions based on information in an embargoed journal article that has not yet become public. At last one legal expert has concluded just that: "The journalist has no duty to the shareholders of the corporation but does have a duty not to use the newspaper's nonpublic information for personal stock trades."[37] Indeed, the journals have tried to warn journalists about this. The American Association for the Advancement of Science, for example, has the following notice in the password-protected section of its EurekAlert! Web site, which gives journalists embargoed access to press releases and articles from *Science:* "Access to this embargoed information is restricted to credentialed journalists, who must disclose any financial conflicts. Anyone using embargoed *Science* information to deal in stocks or securities may be guilty of insider trading, as defined by the U.S. Securities Exchange Act of 1934. This advanced information is intended for individual use, and is not to be distributed to others." Because *Nature* is based in London, its warning is a bit different: "Warning: This document, and the papers to which it refers, may contain information that is price sensitive (as legally defined, for example, in the UK Criminal Justice

Act 1993 Part V) with respect to publicly quoted companies. Anyone dealing in securities using information contained in this document, or in advance copies of a Nature Journal's content, may be guilty of insider trading under the US Securities Exchange Act of 1934."

However, even if science journalists are not using embargoed information for their own financial benefit (and there is no indication that they have been), they nevertheless could be the unwitting source of information for other inside traders. That's because science journalists usually need to get comments from other researchers about a given journal article. Many researchers will decline to expound on a paper if they haven't read it, so science journalists often share their embargoed copies of articles with outside sources for them to read and then offer comment.[38] The journals recognize and tolerate this practice. For example, *Nature* tells journalists: "Solely for the purpose of soliciting informed comment on *Nature* papers, you may show relevant parts of this document, and the papers to which it refers, to independent specialists—but you must ensure in advance that they understand and accept *Nature*'s embargo conditions." This news-gathering process, although it helps journalists produce better coverage of the journal article, may also allow word of financially important research to leak out to some researchers, government officials, analysts, and other experts—and from them, to their friends, relatives, and coworkers.

The embargo offers another, perfectly legal, way for nonjournalists to profit from research findings: find out through nonembargoed sources what the media will report about the journal's findings, and make the right investment before the onslaught of news once the embargo lifts.

Stock analysts do this all the time. Unlike company officials and journalists, who have a duty to keep confidential information confidential, stock analysts usually do not. If they determine that a journal is about to publish a study that could help or hurt a company's stock, they generally are free to act on that conclusion, such as by advising clients to buy or sell stock.

There is substantial evidence that this early trading takes place, and that it has for decades. In 1981, stock in the pharmaceutical firm Merck and Company rose by more than four dollars a share after the *New England Journal of Medicine* provided journalists with embargoed copies of a study showing promising results for a Merck heart attack drug. Wall Street analysts had access to the embargoed information either by being on the journal's mailing list themselves or through contacts with journalists. Jerry Bishop of the *Wall Street Journal* concluded that "this is an intolerable situation for it undermines the concept of a release time. Should it continue, science writers may be forced to begin ignoring release times on journals altogether." In response, in 1982, the *New England Journal of Medicine* started limiting the mailings of embargoed materials to journalists who would promise in writing that they would adhere to the release time. Nevertheless, one medical journalist noted in 1982 that Wall Street securities analysts were increasingly attending medical conferences to glean insights about the prospects for new medications that they could use to advise their clients, long before studies on the medications would be published in a medical journal.[39]

In 1995, stock prices of the biotechnology firm Amgen surged when a stock analyst was able to deduce from journalists' inquiries that the company was about to publish a study in *Science* describing a hormone with the potential ability to help patients lose weight. "Undeterred by press embargoes, investment analysts say they will jump at the chance to pass along advance information that will benefit investors," a news article in the journal concluded. Two years later, trading in the stock of Geron Corporation surged to forty times its usual volume in the hours before the expiration of an embargo on a report in *Science* that researchers supported by the company had been able to grow human embryo cells in the laboratory, a possible precursor toward development of tissues that could be transplanted into humans. "Speculators had somehow gotten wind of the news to come—enabling some to more than double their investments before the media around the country trumpeted the scientific advance," the *Los Angeles Times* reported. In 2002, there was a surge in trading in the stock of Wyeth in the days before the expiration of an embargo on a federal study, published in the *Journal of the American Medical Association*, that documented serious side effects in a hormone-replacement drug sold by the company.[40]

Because of the potential profitability of investing in biomedical companies, stock analysts have developed a variety of techniques, some illegal or bordering on it, for ferreting out information about research. Analysts have gone undercover as research subjects in clinical trials or have posed as physicians to get firsthand information about the progress of the trials.[41] In this context, it is unsurprising that analysts now flock to scientific meetings in search of insights about such research so they can piece together every bit of information they can and in fact are said to often outnumber journalists at such meetings. At least one pretended to be a physician in order to get into a biomedical meeting that barred financial analysts.[42]

And the analysts also find ways to take advantage of embargoes imposed on news coverage of scientific conferences. The most notorious example is perhaps the annual meeting of the American Society of Clinical Oncology, at which biomedical researchers present findings about cancer treatments. Before 2003, the society mailed its members a thick book in advance of the conference containing abstracts of presentations to be made during the event; members were not supposed to divulge the abstracts to anyone else. The society also offered a database of the abstracts on its Web site, with access limited to members of the society, about a month before the meeting. The database was not open to the public—including stock analysts—until the start of the conference, and journalists had to agree to an embargo on media reports about the research, even if the reporters got the information from a third party. Nevertheless, in what has become known as the "ASCO effect," some of the research findings appeared to have filtered out early anyway, because stock of the sponsoring companies had risen (or fallen, depending on how the research turned out) in the days before the start of the conference. Despite the society's efforts to restrict access to the abstracts, it seems likely that some members of the society either passed information to stock analysts or gave the analysts access to the database.[43]

But the analysts don't have to use purloined passwords to profit from embargoes. All they have to do is sign up for a subscription to certain journals. The *Journal of the American Medical Association,* for example, allows anyone to subscribe, and buying a subscription does not obligate one to keep the journal's contents confidential until the embargo expires. But the arrangement also allows analysts who subscribe to make educated guesses about the science news that will be in newspapers and on the air.

The *New England Journal of Medicine* also allows anyone to subscribe, and it mails its issues out a week before the embargo date so that its physician subscribers, too, will have the issue in hand when journalists start reporting on it and patients start calling. But analysts are not reluctant to use the information. For example, in August 1999, a stock analyst advised her clients that Amgen's stock was "attractive" despite results of a study about to be published in the journal that could depress sales of one Amgen drug used in kidney dialysis. The analyst told a journalist that she routinely receives her copy of the journal one or two days before the embargo lifts. The journal editors were unmoved. "We are not responsible for the stock market. We are not an arm of the SEC," replied Marcia Angell, then the journal's executive editor.[44]

And though the practice of analysts taking advantage of the few days between the time when they receive a journal in the mail and the time when the embargo lifts may be unseemly, it probably is not illegal, at least according to one official at the Securities and Exchange Commission. Journals such as the *Journal of the American Medical Association* are sent to so many people that it is not insider trading for subscribers to capitalize on the timing of delivery of the journal's issues, according to the official. "People getting the first copy in that fashion aren't getting an improper advantage. Anybody who wants to can subscribe to the *Journal of the American Medical Association*," he said.[45]

Conclusions

Regardless of whether the embargo promotes more accurate journalism (and Chapter 3 casts doubt on whether it does), this chapter documents that the embargo system causes substantial collateral damage. It has an enervating effect on journalistic competition and tends to perpetuate the work of less skilled journalists by constraining more skilled competitors. The embargoed journals tend to draw media attention from nonembargoed journals, skewing the overall menu of science and medical stories to the topics covered by those embargoed journals. The embargo system also creates a secret underground in embargoed information, which stock analysts and others "in the know" can use to their financial benefit. By promoting a breaking-news approach to science by reporters, the embargo may also foster public misunderstanding of the scientific process, and so may undermine the public-policy goals that many scientific societies claim they seek to achieve through the embargo. Given this list of problems caused by the embargo, the next chapter will consider alternatives to the embargo system.

5

The Embargo Should Go

When they're not praising the embargo system, science and medical journalists often bitterly complain that they are its prisoners. For example, Natalie Angier of the *New York Times* claims that the embargo system gives journal editors "a stranglehold on journalistic initiative."[1] From this point of view, editors of journals exert control over journalists; journal editors dictate what information journalists can have access to and when, and journalists are powerless to resist. However, the previous chapters have presented a more nuanced view of embargoes and how they operate: embargoes do exert great influence over what gets covered and how, but the embargo system is hardly a tyranny of journals over journalists. Journalists are enthusiastic participants in the embargo system and act to keep it functioning. In short, if journalists are in a stranglehold, it is a self-inflicted stranglehold—and one that does not serve the public interest.

It need not be this way. Journalists could do end runs around the embargo, if they wanted. "Any decent journalist knows what's in *Nature* next week," said David Whitehouse, science editor for the BBC's Web site. That point is exemplified by Robin McKie, science reporter for the London *Observer*. He writes for a paper that publishes only on Sundays; because of the embargo's timing, journal articles seem like old news by the time his paper publishes. Consequently, McKie refuses embargoed access to journals but nevertheless has broken news of important embargoed studies such as the cloning of Dolly and the results of decoding of the human genetic sequence. Much of his reporting, he says, is based on talks given by researchers at open meetings.[2]

What accounts for the embargo's staying power? One factor, quite clearly, is the extraordinary deference that the scientific and medical establishments receive in society—not just from journalists but also from government, business, and individuals. Few other institutions are given the freedom of action in society that science and medicine have enjoyed. That consideration has its roots in the role played by science and technology in World War II but persists because of the pervasive impact that science and medicine continue to have in our society. Science and medical journalists have not been immune to the immense gravitational forces exerted by science and medicine, and the embargo system is symptomatic of the deference that journalists pay to the scientific and medical establishment.

This journalistic courtesy is clearly illustrated by the historical account in Chapter 2. From the beginnings of modern science and medical reporting in the 1930s and 1940s, journalists were eager to prove their bona fides to scientists and medical researchers so that those researchers would cooperate with the journalists. Journalists emphasized that they sought to be accurate (as the researchers defined accuracy), and asked the researchers to provide advance copies of their papers to facilitate these efforts. Although the scientific and medical establishments at first were slow to respond, eventually researchers and officials saw the advantages of controlling the flow of news about science and medicine; together, journalists and the research establishment forged the social construction that is now known as the embargo system.

The fact that science and medical journalists place themselves under the thumb of science and medicine is underscored by the way in which disputes about embargo violations are judged: each journal arrogates to itself the right to determine whether a journalist has violated the embargo and what the penalty, if any, should be. Some—like EurekAlert!—do consult with journalists, but the decision-making authority remains with the journal. One could imagine an alternative arrangement for ruling on embargo disputes, such as making use of a panel of uninvolved journalists and officials from journals other than the one that alleges the violation.[3] Such an arrangement would place the research establishment and journalists on a more equal footing—which would disrupt the power relationships that both groups have cultivated over decades. Christopher Dornan writes that institutional science has insisted "that science is the rightfully dominant authority over the adequacy of press coverage of any issue to which science contributes. There is a slide from the premise that journalism should be required to get the facts right, to the assertion that these details themselves dictate the form and tone coverage should adopt."[4]

Science Journalism and the Public Interest

All forms of journalism—not just journalism about science and medicine—should serve the public interest. To meet that obligation, journalists and media organizations should help readers and viewers to function intelligently in society, whether as citizens, consumers, or family members. As a recent critique of journalism succinctly put it: "The primary purpose of journalism is to provide citizens with the information they need to be free and self-governing."[5]

To fulfill this obligation to serve the public interest, science and medical journalists should provide readers and viewers with information that they need to understand, and make judgments about, matters of public controversy, such as cloning, environmental issues, and military technology. Science and medical journalists should also aggressively and thoroughly plumb the workings of the scientific and medical establishment, virtually all of which operates on public funds, either directly through grants or indirectly through tax-sheltered institutions such as private universities and foundations.

Supporters of the embargo say that the embargo does promote coverage of important scientific research. But these well-meaning proponents of the embargo system confuse what is good for the scientific and medical establishments with what is good for society. Similarly, journalists erroneously equate what is in their interest with what is in the public interest.

Undoubtedly, the scientific establishment benefits handsomely from the unending torrent of news coverage about research being published in scientific and medical journals. The pattern of news coverage signals to readers and viewers—not to mention lawmakers, business leaders, and others—that science and medicine are important. That impression is vital for the scientific and medical establishments, which depend on largesse from the government, business, and even individuals. Moreover, whether the research being reported is "good news" (for example, drug X is an effective treatment for disease Y) or "bad news" (Z causes cancer), the scientific and medical establishments are always cast in a positive light, as the font of the new finding. Some think that breathless coverage of the latest research might even help steer some individuals toward careers in science or medicine.

Indeed, news coverage of new research invests science and medicine with an air of excitement and importance. Two critics of media coverage of medical reporting, based at Dartmouth Medical School, have found such a symbiosis in the process of news coverage of medical meetings: "Media outlets are in intense competition with each other for a limited audience; meeting organizers need to attract scientists, advertisers, and sponsors; researchers need to show results to advance their careers; and academic institutions need publicity to raise funds," say the two, Lisa Schwartz and Steven Woloshin. "In each case, self-interests are served by being associated with work that is perceived to be new, big, and important."[6]

Their critique also fits media coverage of research journals, which often amounts to little more than highbrow infotainment: What's the latest theory about the extinction of the dinosaurs? What's the newest thing found to cause cancer? Look at the cool photographs from the Hubble Space Telescope! These are the types of subjects that dominate embargo-controlled news reporting about science and medicine.

Journalists and their media organizations—particularly those with daily deadlines, such as newspapers, network television, and Web sites—also benefit from the embargo. The embargo supplies news on a dependable schedule keyed to the production constraints of news organizations: if it's Thursday, it's time for a newspaper article about some paper published in the *New England Journal of Medicine.* The embargo reduces the stress on journalists from having to rush a story into print or on the air, and it also reduces the possibility that the journalist will be scooped by a competitor. The news peg provided by the embargo ("In a paper published today in *Science . . .*") also makes it easier for journalists to convince their editors to run certain research-news stories. This news peg supplies an appearance of sudden and urgent newsworthiness to a research paper that had been conceived, executed, written, reviewed, and rewritten over a period of months, if not years. By keying the news story to the event of journal publication, the embargo system capitalizes on the fact that science and medical journalists and their editors—like other journalists—rely heavily on timeliness as a criterion in defining what is news and what is not. In fact, science and medical journalists are fully aware of the artificiality of the embargo's news peg, but rely on it as a way to get research news into print and on the air.

This is a long-standing problem in journalism about science and medicine. "To write a story saying that 'X' was discovered today is a fiction," Howard Simons, then a science writer for the *Washington Post,* said almost forty years ago. "The today lead is something most of us do because we are still trapped in traditional ideas of newspapering. At a scientific meeting there may be hundreds of papers delivered, all of them important. There is no reason why we shouldn't pick up one of those papers three weeks later and do a story about it. But the traditional light bulb flashes on in our minds and says it's old if it's not hung up like a coat on a news peg."[7] Although the urgent demand for a news peg—even an artificial one—for science and medical news is not new, the embargo perpetuates the problem by giving journalists and their media organizations an unending stream of such pegs, so many that a lazy journalist could write only about journal articles if he or she chose.

Journalists' role in perpetuating and even extending journal embargoes is illustrated by the recent history of embargoes at online journals published by BioMed Central. BioMed Central, like a few other innovative journal publishers, has seized on the Internet as a medium for publishing research journals at lower cost and with wider reach than traditional journals. These new online journals are known as open-access journals, because anyone can read them online for free; the journal publisher instead gets its revenue from charging the researcher who publishes in the journal (and this charge is often covered by whatever government agency has provided the researcher's grant). One of its hallmarks is that it publishes a journal article as soon as peer review and editing are complete, without the lengthy delays that sometimes characterize printed journals.

When BioMed Central started in 2000, it did not offer advance embargoed access to journalists; journalists had to wait to read a journal article until it was publicly posted on BioMed Central's Web site. But journalists ignored its journal articles. "We found that we were not getting coverage by journalists because they need embargoes," notes Grace Baynes, press officer for BioMed Central. "What journalists want is for their news piece to appear on the same day that the journal article went live."[8]

Consequently, in 2003, BioMed Central tweaked its editorial processes so it could start offering journalists a brief period of embargoed access—a few days to at most a week. Generally, the journalists have embargoed access to a version of the paper that is complete but not in its final format. But sometimes BioMed Central does ask authors of a paper for permission to delay public release for a day or two, to provide a window for an embargo. The authors usually agree, Baynes says. "Invariably, they're very excited about the fact that we want to promote the paper, and they're happy to wait a day or two." Since the change, BioMed Central has seen a marked increase in press coverage. Baynes believes that BioMed Central benefits from the press coverage. The publicity about a BioMed Central article may lead a scholar to read the journal and perhaps submit a paper of his or her own, she says. And the authors whose paper was publicized may well want to submit future work to BioMed Central. "It's exciting for authors. We really want our authors to publish with us again."[9]

The fact that journalists and science benefit from the embargo does not mean that the public benefits from it as well. Indeed, the embargo works against the public interest in many ways. One is in how the embargo steers journalists away from covering science and medicine as institutions with messy problems, such as fraud, mistreatment of human subjects, failed research, and misplaced priorities. Journalists who are chasing after the latest embargoed journal article do not have time to investigate the workings of science and medicine in this way. "Sometimes I think the reason science writers do so little big-scale investigative reporting is that they exhaust themselves just trying to get the stories right on a day-to-day basis," states Deborah Blum, a Pulitzer Prize–winning science journalist. "The nature of our job provides little time to burrow in."[10]

Indeed, the central problem with embargoes—and the reason that the embargo system should be eliminated—is that embargoes are a distraction for journalists and their media organizations, which diverts them from covering what really matters. "To survive, reporters become dependent on the daily cascade of embargoed research papers, e-mailed press releases, university tip sheets, and conference abstracts," says Robert Lee Hotz, a science reporter at the *Los Angeles Times.* "The goal of all of us ought to be to try to get around embargoes and packaged science journalism by finding new ways to get our collective noses under the tent," according to Cristine Russell, former science and medical reporter for the *Washington Star* and the *Washington Post.* "I think that we should not have such herd journalism. . . . I am very concerned that we get very lazy and complacent by sitting around waiting for the journals to deliver us the news that something's happening in science. . . . I think people should get out and cover science."[11]

Of course, no one forces journalists to report embargoed stories. In part, the embargo's focus on late-breaking research appeals to the personal and professional interests of many science and medical journalists, who entered the field to report about the wonders of research—not the bare-knuckles reality of the research world. "Most science reporters tend to behave rather like sports writers: they have chosen their topic out of love for it," the sociologist Dorothy Nelkin has observed. "Some reporters can be seduced by the wonder," one journalist points out. "Traditional science journalists have focused on interesting research," according to another, "covering science's underbelly only when rumors of scientific malfeasance emerged from a scientist whistleblower or other easily available source or when a credit spat erupted."[12]

The embargo system also works against the public interest in the way that it misleads the public about science and medicine. The embargo creates a torrent of news that draws excessive public attention to most research. Put simply, journalists should ignore most of the journal articles that they now cover so energetically. Most journal articles, whether in medical or nonmedical journals, are but single dots in the pointillist enterprise that is the scientific method—but the breathless coverage catalyzed by the embargo system often gives the impression that each week's paper is a major breakthrough. "Single studies represent just one link in the chain of scientific understanding and are frequently not conclusive," four nutrition experts have argued. "For example, they may be in conflict with the current state of knowledge, too small to generalize stable results contrary to the consensus of opinion, or not generalizable. . . . Findings from single studies are often just one of many pieces in an intricate puzzle. The true picture begins to emerge only as the many pieces are carefully put together." Two other critics say: "The 'breakthrough syndrome' overemphasizes good news—for example, a promising but very small study of Alzheimer disease or an animal experiment without clear clinical importance. Scary epidemiology can be sensationalized, as occurred when media reports about pancreatic cancer and coffee were substantially less cautious than the discussion section and conclusion of the journal articles." Journalists pay much less attention to later studies that play down the findings.[13]

The best science and medical journalists recognize the inherent uncertainty in any piece of research. "It takes repeated observations or experiments, usually attacking the mystery from different angles with results all pointing to the same answer, before honest researchers begin to believe that they actually understand something new," says Boyce Rensberger, a former science journalist for the *Washington Post* and now the director of a midcareer science-journalism fellowship program at the Massachusetts Institute of Technology.[14]

A third way in which embargoes oppose the public interest is in undermining the usual and important competitiveness among journalists and their media organizations. Society needs its journalists to be vibrant and assertive and enterprising in their efforts, in order for journalism to have the ability to thoroughly investigate the functioning of society. Competition promotes that vibrancy.

These days, competition among journalists suffers from a bit of a bad rap. The notion of "competition" seems to evoke the quest among some media outlets to run the most footage of Monica Lewinsky or O. J. Simpson or John F. Kennedy Jr. or Martha Stewart. Such competition, many will agree, does not particularly serve the public interest. But other forms of competition do serve the public interest. Journalists of all stripes—and their media organizations—should spend time, energy, and money poking into the functioning of society, and they should compete with one another to do so.

Certainly, the best science and medical journalists already do this, despite the blandishments of the embargo. Deborah Blum won her Pulitzer Prize for in-depth reporting for the *Sacramento Bee* on the conduct of primate research. Laurie Garrett likewise won a Pulitzer in 1996 for her reporting for *Newsday* on the Ebola virus. John Crewdson, a reporter for the *Chicago Tribune,* scornfully rejects the title of science journalist, but he has aggressively investigated allegations of wrongdoing in AIDS research at the National Institutes of Health.

However, among science and medical journalists, the embargo promotes passivity and reactiveness, rather than an enterprising, proactive stance in reporting. "My biggest beef with the embargo system is that it sucks energy away from the healthy competitive drive that applies in most other news arenas—the news-breaking hunger for the scoop," says Bob Beale, an Australian science journalist.[15] The embargo system steers the attention of all participating science and medical journalists to a small set of scholarly journals. All science and medical journalists plumb these journals for news, and, perhaps more important, all participating journalists know that their competitors are also plumbing that same set of journals. This behavior means that all journalists are much more likely to cover the same papers, reducing the risk that any given journalist will cause difficulties for others by covering a paper in some other journal. In general, society should be promoting more, and better, competition among journalists to uncover information important to individuals and society.

The issue of competition relates to one of the common defenses of the embargo—that it provides a "level playing field" for all journalists who seek to report on a scientific finding. When they invoke this rationale, journal editors never explain how a level playing field serves the public interest. In fact, it does not: reporters who are talented and fast on their feet should be able to distribute their news coverage as soon as they are able, even if less talented journalists are left behind. This process, Darwinian as it may seem, promotes excellence by journalists and their media organizations. Scientific and medical journals could provide a sufficiently level playing field by simply releasing their information to all journalists at the same time, without favoring one outlet over another.

In journalism, competitiveness arises from independence. That is, if journalists are independent of one another and of their sources, their competitive juices lead them to look for stories that are important, and they will seek to report those stories first and more fully than other journalists. Indeed, in their recent critique of U.S. journalism, Bill Kovach and Tom Rostenstiel argue that journalists should exhibit an "independence of mind and spirit" relative to their sources.[16] Journalists should not wait for an authority figure—in this case, publishers of scientific journals—to identify the "news" for them and provide it on a silver platter. Rather, journalists should be on the prowl for stories that investigate and illuminate the workings of government and society—stories that institutions such as science and medicine may well prefer journalists avoid.

Pack journalism, of course, is hardly unique on the science and medical beats. Indeed, political journalists also exhibit noncompetitive behaviors, as documented most famously by Timothy Crouse in his account of pack reporting in the 1972 presidential campaign. And news sources use embargoes on topics other than science and medicine, often for justifiable reasons. For example, the Commerce Department routinely gives financial journalists access to trade statistics one-half hour before they are publicly released. The journalists are required to remain together in a locked office but can start writing stories for release when the embargo lifts. Also, congressional committees often give journalists advance copies of the prepared texts of witnesses at scheduled hearings for the same purpose, and companies occasionally provide embargoed access to product announcements.[17]

Both science and medical journalists and their nonspecialist colleagues allow themselves to be vulnerable to bureaucracies that understand their needs. "In the modern era, the institutions that journalists cover most intently have learned to feed reporters a steady diet of news, or what passes for news, just to keep them busy," two senior editors of the *Washington Post* recently observed. "It is difficult to break out of the day-to-day routines of journalism, avoid the ordinary distractions and dig for hidden news. The instinct to do this doesn't come naturally to most reporters."[18]

The problem could worsen; in the face of the increasing volume of embargoed information available to them through avenues such as EurekAlert! science and medical journalists may find themselves with less and less time to conduct independent reporting, such as in-depth looks at the functioning of the scientific establishment. This, too, is not a problem unique to journalism about science and medicine. Kovach and Rostenstiel issue a similar warning, albeit about reliance on computerized databases in news gathering: "As journalists spend more time trying to synthesize the ever-growing stream of data pouring in through the new portals of information, the risk is they can become more passive, more receivers than gatherers."[19]

Journalists' ethical obligation to focus on serving the public interest means that they should not allow their media organizations to be placed in the service of the interests of others. As this book has documented at numerous points, journal publishers and scientists use the publicity produced by the embargo system for their own ends, and private investors who can gain inside access to embargoed information stand to benefit financially, while members of the public have to wait to read about a study in the newspaper or to hear about it on television before they can take advantage of the information.

Accuracy

The trump card for embargo supporters is accuracy: news about science and medicine is so difficult to research and write, goes this argument, that journalists need time to do the job correctly—or the public could be harmed by inaccurate reporting. "It generally is conceded that this news is so important that it transcends other principles of the media business," according to Peter Gorner of the *Chicago Tribune.* "Nowhere else in journalism is being right considered more important than being first." Susan Turner-Lowe, former director of public affairs at the National Academy of Sciences, describes it this way: "Journalists have traded accuracy for scoops."[20] Among many proponents of embargoes, being critical of embargoes therefore is tantamount to supporting erroneous reporting.

But the fact is that many other journalists work effectively without embargoes, day after day. Consider the complexity and implications of other stories covered by journalists who do not specialize in science and medicine: the latest Supreme Court decision, a tax bill passed by Congress, a massive airplane accident, and others. Each of these stories rivals many science and medical stories in technical complexity, the difficulty that journalists may have in reaching expert sources for comment, and the impact on readers or listeners if inaccurate information is reported. Yet reporters uncomplainingly cover these and a myriad of other stories without the helping hand of an embargo. Even *Nature*'s Peter Wrobel concedes that the embargo is not essential for good coverage of science and medicine. "It doesn't require five or six days, or even three, to write most stories," he said. Indeed, Alexandra Witze, a science reporter for the *Dallas Morning News,* says that the accuracy rationale for journal embargoes is "insulting" to science journalists. "It assumes that we are incapable of doing our job as journalists in any other field are."[21]

However, views such as hers are scarce among science and medical journalists, who agree with journal publishers that embargoes serve the public interest because embargoed advance access to scholarly journals promotes accurate, orderly journalism about science and medicine. As this book documents, this is not necessarily the case. Journalists who operate by a learned set of professional norms and practices are likely to make the same mistakes in a story whether they have a day or a week to prepare it. Moreover, an individual reporter may not use all the additional time that the embargo provides. With an embargo of several days, the reporter may work on the embargoed story in bits and pieces, fitting that story around other stories that the reporter is covering.

Journalists do have an ethical obligation to society to be accurate, but accuracy is more than the technical accuracy of figures and scientific terms. Taken as a whole, science reporting should provide an accurate picture of scientific and medical research, particularly in areas of personal importance to members of the public, such as health issues. The embargo arrangement encourages pack reporting of research from a few selected journals regardless of whether the research is truly important or definitive.

Furthermore, the concept of accuracy is slippery, as Chapter 4 shows. The evaluation of news accuracy can well be regarded as a social construction, that is, as a reflection of the interactions between journalists and scientists. In short, scientists dictate the definition of accuracy that the embargo is designed to promote, so it is unsurprising that both scientists and journalists insist that the embargo promotes more accurate journalism. Other conceptualizations of accuracy—such as the accuracy of the weekly rat-a-tat-tat of minor findings related to health, hazards, and other concerns to readers—remain largely ignored by embargo participants.

It also bears noting that although the editors of scholarly journals that operate under embargoes voice support for the practice, it clearly does not enjoy total support in academe. Some scholarly societies—such as the American Geophysical Union and others—see no need for an embargo or believe that it does not operate in the public interest. Individual scholars also criticize the embargo arrangement. Some characterize it as a type of collusion that interferes with the stated purpose of scholarly communication: "Science is supposed to progress through rapid communication of results among scientists, but the embargo system is a barrier to this free exchange of information. One can understand that publishers do not want to feed the public with incomplete and inaccurate information but other scientists in the academy would have liked to enjoy the same kind of privilege extended to the media."[22]

Impact of the Internet

In the short run, the Internet has probably bolstered the embargo system, particularly because the World Wide Web and electronic mail have provided new tools for distributing embargoed articles to journalists.[23] EurekAlert! in particular has been a resounding success story for embargo proponents, so much so that it has spawned imitators such as *Nature*'s press Web site and AlphaGalileo.

But in the long run, online communications will probably undermine embargoes on news about science and medicine. One reason is the very phenomenon noted in the previous paragraph as an argument that the Internet has initially bolstered the embargo: the ease with which the Internet can connect journal publishers with a worldwide cadre of journalists. More and more science and medical journalists, around the globe, are participating in embargoes sponsored by journals in the United States and Britain. Many of these journalists may not be as heavily invested in the embargo system and therefore are more likely to jump the gun when an important paper comes along.

Put more simply: Decades ago, when journal embargoes began, science and medical journalists constituted a handful of reporters who worked alongside one another constantly, at various scientific conferences, meetings, and events. A science or medical journalist who broke an embargo knew that he (and back then, a science or medical journalist almost always was a "he") would face angry questions from fellow journalists the next time he encountered them. By contrast, the journalists who today participate in the embargo are scattered around the world; many are unlikely to ever have to face the journalists in the United States and Britain who would be most discomfited by an embargo violation. Or from the standpoint of the social construction of news, the widening of the pool of potential embargo participants made possible by the Internet's global reach may seriously erode the social bonds that maintain the embargo. As the embargo comes to include more journalists around the world, the participants are likely to include individuals who have not been socialized to honor embargoes and who would be immune to social pressures from other embargo participants to conform to embargo rules.

The Internet will also weaken the embargo because it is transforming the process of scientific communication itself. Most traditional journals now offer online access to their articles, with the articles often posted long before the printed journal arrives in a scholar's mailbox. Some journals have gone a step further, by publishing some or all of their articles online before they are published in print. The *New England Journal of Medicine,* for example, has posted on its Web site certain articles that, in the opinion of its editors, were in the public interest for rapid dissemination. Although the New England journal generally restricts online access to paid subscribers of the journal, anyone could read or download these "early release" articles.[24] *Science* and *Nature* have also begun to post selected journal articles online, after they have completed peer review and editing but before they appear in print.

Increasingly, early online publication appears to be becoming the norm for at least some journal publishers. The *Proceedings of the National Academy of Sciences of the United States,* for example, publishes all articles online before they appear in print, as much as five weeks later. The American Chemical Society also publishes online some of the articles issued in its twenty-seven journals as soon as they are ready for publication, even if they have not yet appeared in a printed publication. The articles are posted online after they have been peer-reviewed, copyedited, and checked by the author. Embargoes for these articles are much shorter than embargoes for printed journals.[25]

The Internet is also changing other aspects of communication among scientists. One example is the departmental seminar, often conducted in the late afternoon, perhaps along with punch and cookies, at which a local researcher or a visiting scholar offers a brief presentation on recent research. Increasingly, departments are transmitting live video of these seminars across the Web, using a technology known as streaming video, to anyone who cares to watch and listen. The National Institutes of Health, for example, Webcasts many such seminars each week (http://videocast.nih.gov). The Multi-University/Research Laboratory (http://murl.microsoft.com)—a joint effort of computer scientists at six universities or corporations—similarly transmits live seminars in that field.

Although it is impractical for most science journalists to cover departmental seminars in person, nothing stops journalists from covering these broadcasts as news, or watching them as a way to develop leads on future research developments. There is little indication that journalists are using Webcasts as a reporting tool to date, but journalists could use the fact of such a live Webcast (and, more compellingly, the existence of a freely available online copy) as evidence that a particular research finding is in the public domain, vacating any embargo. (Interestingly, the *New England Journal of Medicine*—one of the most staunch foes of online prior publication—has ruled that it will not disqualify from publication an audio recording of a presentation at a medical conference along with "selected slides from the presentation.")[26]

Scientists are also using the Web to archive and distribute preprints of their papers. As noted in Chapter 1, scholars have long circulated preprints, or papers that have not yet been submitted for publication in a journal, to colleagues for their comments and suggestions. With the advent of the Web, scholarly societies and even individual scholars have created databases on which authors can deposit electronic copies of their papers. Few journalists use the Web sites to plumb for news.

One who does is Tom Siegfried, science editor of the *Dallas Morning News.* "There's plenty of stuff to report out there before they appear in journals," he says. Every night, he says, he checks physics preprint servers, because the latest research is usually reported there first. "In physics nowadays the journals have become increasingly irrelevant," he contends, with their role largely limited to serving as the archival copies of important papers and for proving records for tenure. A journalist has to use judgment in deciding whether a paper is worth covering, but many nuggets are there nonetheless, he notes.[27]

Even aside from these new forms of online scholarly communication, the Internet weakens the embargo system by providing new routes for embargoed information to leak into the public sphere. "A lot of my reporting is Web surfing," says Alexandra Witze, also of the *Dallas Morning News.* "Journals are one tool among many. I think the problem is that many journalists focus on them as the only tool."[28]

Internet mailing lists, newsgroups, Web sites, and Web logs all provide venues for interested nonjournalists to swap information about research news. Journals publish tables of contents of upcoming issues. Scholarly societies post abstracts of papers to be presented at their scientific meetings. Support groups for individuals suffering from particular diseases and their families, for example, are likely to operate online information sources that discuss clinical trials of treatments for the disease in question. Attentive journalists monitoring these Web sites and mailing lists could well glean information that could lead to stories outside the embargo.

Such a surfeit of information will increasingly create situations in which science and medical journalists decide that online information has vacated the embargo on a specific journal article. Indeed, this was one of the justifications made by the *Detroit Free Press* and journalists at other media organizations who prematurely reported findings on a study of hormone-replacement therapy in 2002; the findings, they said, were already being discussed online: "Women's health sites on the Internet were buzzing about what the study said and what women should do."[29]

Another way in which the Internet weakens the embargo is by amplifying the effects of an embargo violation by one media organization. Prior to the Internet, when a minor organization broke the embargo, other media organizations might not immediately realize what had happened. Most journalists might well continue to adhere to the embargo, either out of ignorance or out of a belief that the breach was a small one. Today, however, anyone can read freely available media Web sites published anywhere around the world. Embargoed information published on the Web site of one media organization can attract enormous attention and can start a major stampede among journalists to disregard the embargo. As Chapter 2 recounts, some journalists anticipated such a development when the news of the cloned sheep Dolly was distributed to science journalists under embargo. The predictions were proven true when media organizations in Italy and Great Britain reported the research before the embargo time. Under such circumstances, journal publishers are fooling themselves if they believe that they can use embargoes to control the timing of the release of major scientific news.

A World without Embargoes

The embargo system should be replaced with full and open disclosure of research results as soon as they are ready for public consumption, which generally would mean as soon as peer review is complete. Once a scholarly paper has been accepted by a journal, scientists and their institutions should be free to tell the world about it, and journalists should be free to report on it if they deem it newsworthy. As many have already begun to do, the journal in question could make the accepted paper available to its subscribers online, so that the subscribers could consult the full text of the paper for themselves. Journalists would be freed of the perceived tyranny of the embargo, and they would have newfound time to visit scientists in laboratories and troll for investigative stories rather than leafing through press releases and password-protected Web sites in search of what the competition is probably going to report.

This is emphatically not to suggest that science and medical journalists should break embargoes. To the contrary, journalists have both an ethical and a legal duty to abide by agreements with their sources, including embargo agreements. If a journalist obtains information under an embargo, that journalist is ethically bound to honor that embargo, just as the journalist would be ethically bound to, for example, withhold the name of a source if the journalist agreed that the source would be unidentified.

But although journalists are ethically bound to honor embargoes to which they have agreed, they are not ethically required to continue to agree to embargoes. Continuing the parallel to anonymous sources, many media organizations have established policies that govern the conditions under which they will grant anonymity to a source.[30] But the fact that a media organization has had a policy for granting anonymity in the past does not mean that it will always grant that anonymity to all sources in the future. Similarly, the fact that science and medical journalists have used embargoed information in the past, and have respected those embargoes, does not mean that those journalists or their media organizations must continue to agree to embargoes in the future.

In short, science and medical journalists, and their media organizations, should terminate their current embargo relationships with journal publishers and stop accepting embargoed information from them. Scientific societies and journal publishers should stop distributing information under embargo. Government research agencies and foundations should stop supporting the embargo, which provides a few with privileged early access to taxpayer-financed research. Universities, which cast themselves as champions of free expression, should oppose embargoes on their faculty members' research, rather than seeking to hitch their own publicity machines to the journals'. It is time for science and medical journalists to break out of their dependence on journals as a source of science news, and it is time for scholarly societies to stop trying to shape the flow of news in a way that suits their own political ends.

How would science and medical journalism be different if embargoes went away? Some have a dire vision of a feeding frenzy each week, as journalists clamber for whatever edge that can find over fellow journalists in finding out about the latest research. Of course, such incidents happen even now, with embargoes. If embargoes were eliminated, the more likely outcome is that the popular press would simply ignore much of the new research that seems so urgent today because of the false patina of newsworthiness that is conveyed by the embargo. Without the embargo's bogus news peg, editors and producers would judge many of those research reports to be less compelling. Knowing that their competitors at other newspapers and broadcast outlets would act the same, they would feel less competitive pressure to publish the "latest" research each week. That could well translate into an overall decline in news coverage about journal articles, particularly about marginal journal articles that advance science only incrementally and ambiguously.

When journalists and their organizations did choose to cover a certain paper, they would not have to release their work at an arbitrary time set by an outside source. They would have total control over the timing of the report; whenever a news organization judged that its coverage was ready to run, it could run that news coverage. In some cases, that might be in as little as a day. In other cases, a news organization could take several days, a week, or even more. News organizations make this type of news judgment all the time, balancing competing factors such as the completeness of the story and editors' confidence in its accuracy against concerns such as staying ahead of (or at least abreast of) the competition.

Indeed, editors would play a major role in the death of the embargo. Science and medical reporters are not solely to blame for the embargo; they adhere to the embargo so they can stay out of hot water with their editors, who (the reporters believe) have little understanding of, or appreciation for, the importance of news about science and medicine. Science and medical journalists believe that it is at least potentially problematic for them not to have reported on a scientific paper that a competing news organization had elected to cover—regardless of how important that paper really was, and regardless of the fact that covering that paper might divert time and energy from covering more important topics.

Under such conditions, it is hardly surprising that reporters would agree to an arrangement that circumscribes competition as sharply as journal embargoes do, and in many cases asking individual journalists to abandon embargoes would be asking them to commit professional suicide. Rather, moving science and medical journalism away from an extreme reliance on embargoes would first require a shift in the attitudes of editors who establish the work expectations for the science journalists who participate in embargoes. Editors should understand the embargo and how it shapes the way in which their news organizations cover science and medicine—and, perhaps most important, what the embargo prevents their news organizations from covering. Some editors then may decide that their public service obligations require their news organizations to move away from embargo-assisted journalism.

Of course, even if some news organizations were to reduce their coverage of journals, other news organizations might decide to continue to aggressively cover research findings. It seems likely, for example, that the Associated Press and the *New York Times* would seek to continue to report, with great promptness, on what they each judge to be the most important articles published in elite journals each week. Even without embargoed access, these media organizations would likely seek to develop reporting techniques and approaches that would enable them to get access to journals as soon as they were published, such as dispatching a weekly courier to pick up a copy of each journal from the journal publisher's office.

Newspapers, Web sites, and broadcasters could take advantage of this breaking-news reporting, such as by running AP stories and articles distributed by the *New York Times*'s syndicate. Many, of course, already run such articles. But a better development would be for media organizations to ignore these breaking-news reports for all but the most important research papers. As this book has explored, breaking-news coverage of scientific and medical research misleads the public about science and gives them false hope (or unjustifiably dashes hopes) about new medical treatments. Media organizations would work more in the public interest if they ignored most research findings and instead delivered news about science and medicine that really mattered.

In the cases of truly major and significant research findings without an embargo, of course, science and medical journalists would probably have to work fast. Particularly with a major story, the journalists would have to scramble to get access to both the authors of the article and outside experts who could comment on the study's findings and implications. The journalists would have to quickly grasp the study and figure out a way to translate it into a newspaper story or television report that the general public would understand. And they would have to do all of this quickly—perhaps in as little as a single working day.

This would be a manageable challenge for skilled journalists and media organizations that provide them with the resources that they need, but journalists and media organizations that are not up to the task would produce second-rate coverage. Over time, it would become clear—to scientists, readers, viewers, and other journalists—which was which. Less capable journalists would no longer be sheltered from competition by the embargo's artificially level playing field. Some might be reassigned or disciplined, but the quality of science and medical journalism would rise over time.

It is a rough-and-tumble vision of the journalistic future, one lacking the gentility that now pervades journalism about science and medicine. But the public interest, not the interest of the scientific and medical establishments, should be the uppermost concern of science and medical journalists—and, in fact, of institutional science and medicine. The embargo should go.

Appendix:
Methodological Notes

Most chapters of this book refer to a content analysis of science coverage by U.S. daily newspapers and network television. Rather than burden the text with methodological details, they are presented here for the interested reader.

The content analysis described toward the end of Chapter 1 focused on the breaking-news coverage of the *Journal of the American Medical Association, Nature,* the *New England Journal of Medicine,* and *Science.* Coverage was viewed from twenty-five daily newspapers, the Associated Press, and the evening news broadcasts of the three major U.S. networks from June 1997 through May 1998. Also analyzed were the weekly press releases produced during that period by the publishers of the *Journal of the American Medical Association, Nature,* and *Science* and editorials published in the *New England Journal of Medicine.*

The newspapers included in the content analysis were determined by preparing a list of candidate newspapers, representing all newspapers whose full texts were available to me at the time (May 1997). I submitted this list to officials at the publishers of the four journals in May 1997; they then identified which of those newspapers were recipients of the embargoed materials distributed by that journal. This procedure eliminated from consideration any newspaper that could not conceivably have been directly affected by a particular journal's embargoed materials by virtue of the fact that the newspaper did not receive those materials. The content analysis covered twelve months starting in June 1997, to make sure that the data represented newspapers that were participating in journal embargoes during the content-analysis period. The newspapers were the *Anchorage Daily News, Arizona Republic, Atlanta Journal-Constitution, Baltimore Sun, Boston Globe, Charlotte Observer, Chicago Tribune, Christian Science Monitor, Columbus Dispatch, Los Angeles Times, Miami Herald, New Orleans Times-Picayune, Newsday, New York Times, Orlando Sentinel, Palm*

Beach Post, Philadelphia Inquirer, Pittsburgh Post-Gazette, Richmond Times-Dispatch, Sacramento Bee, San Francisco Chronicle, San Jose Mercury News, St. Louis Post-Dispatch, USA Today, and *Washington Post.*

To gather stories that covered the four journals, databases in Dialog and Lexis-Nexis for each newspaper and network and the Associated Press were searched for all articles containing the titles of each journal from June 1997 through May 1998. Because such a search could return articles that were not of interest—such as those that used the word *science* or *nature* in contexts other than the name of a journal—all articles were initially gathered in a format, known as "keyword in context," which includes only a small portion of the article's text, including the search's target word. I examined each such excerpt and identified the articles that did quote one of the four journals of interest. Those articles were retrieved for content analysis.

Because the aim of this research project was to examine the routine functioning of the embargo system in breaking-news coverage, I retrieved only those articles that were published on the release date for each journal—Wednesday for the *Journal of the American Medical Association,* Thursday for the *New England Journal of Medicine* and *Nature,* and Friday for *Science.* Broadcast stories were limited to the evenings before these times, to correspond with the advantage given to broadcasters in the embargo release times. Also, stories from the Associated Press included articles that were moved the day prior to the journal release date because those articles would be available for use by the newspaper on the release date. News coverage of journal articles for which the journal had lifted or shortened the embargo was also excluded, because those stories represented exceptions to routine coverage. Editorials, letters to the editor, and other items published by the newspapers that were not breaking-news stories were also excluded.

For the comparison of news coverage of embargoed and unembargoed journals discussed in Chapter 4, the same search strategy was executed for newspaper and television coverage, but substituting the names of the American Geophysical Union's flagship journals, *Geophysical Research Letters* and the *Journal of Geophysical Research.*

Notes

Chapter 1: An Overview of News about Science and Medicine

1. Mark Fishman, *Manufacturing the News* (Austin: University of Texas Press, 1980).

2. National Association of Science Writers, *Membership Directory and Resource Guide* (Hedgesville, W.Va.: National Association of Science Writers, 2005), inside back cover; e-mail message to author from Melinda Voss, executive director of the Association of Health Care Journalists, April 29, 2004.

3. Project for Excellence in Journalism, "The State of the News Media 2004" (2004), http://www.stateofthenewsmedia.org/2004/ (accessed January 9, 2006).

4. Larriston Communications, *Guide to Medical and Science News Media* (New York: Larriston Communications, 1990).

5. Bruce V. Lewenstein, "A Survey of Activities in Public Communication of Science and Technology in the United States," in *When Science Becomes Culture: World Survey of Scientific Culture, Proceedings I,* edited by Bernard Schiele (Boucherville, Quebec: University of Ottawa Press, 1994), 150.

6. Terrence J. Sejnowski, "Tap into Science 24–7," *Science* (August 1, 2003): 601; Ira Flatow, "Presentation by Ira Flatow to the National Science Board, Irvine, California, February 3, 2000," http://www.nsf.gov/nsb/meetings/2000/feb/iflatow.doc (accessed August 15, 2003); Project for Excellence in Journalism, "State of the Media."

7. Project for Excellence in Journalism, "Gambling with the Future," *Columbia Journalism Review* (November–December 2001): 1–16 (supplement); Project for Excellence in Journalism, "On the Road to Irrelevance," *Columbia Journalism Review* (November–December 2002): 89–104.

8. Andrew Holtz, "Frustrations on the Frontline of the Health Beat," *Nieman Reports* (Spring 2003): 7–9; Gail Porter, U.S. Department of Energy, and National

Institute of Standards and Technology, *Communicating the Future Best Practices for Communication of Science and Technology to the Public: Conference Proceedings, March 6–8, 2002* (Washington, D.C.: U.S. Government Printing Office, 2002).

9. Project for Excellence in Journalism, "State of the Media"; Guido H. Stempel III, "Topic and Story Choice of Five Network Newscasts," *Journalism Quarterly* 65 (1988): 750–52; Hyuhn-Suhck Bae, "Product Differentiation in National TV Newscasts: A Comparison of the Cable All-News Networks and the Broadcast Networks," *Journal of Broadcasting and Electronic Media* 44, no. 1 (2000): 62–67.

10. Lewenstein, "Survey of Activities," 148; Project for Excellence in Journalism, "State of the Media"; Flatow, "Presentation by Flatow."

11. Project for Excellence in Journalism, "State of the Media."

12. Lewenstein, "A Survey of Activities."

13. Marianne G. Pellechia, "Trends in Science Coverage: A Content Analysis of Three U.S. Newspapers," *Public Understanding of Science* 6 (1997): 49–68.

14. Edna F. Einsiedel, "Framing Science and Technology in the Canadian Press," *Public Understanding of Science* 1, no. 1 (1992): 89–101.

15. Lewenstein, "A Survey of Activities"; Jim Dawson, "The Devolution of a Science Page," *Nieman Reports* (Fall 2002): 16–17.

16. Jane B. Singer, Martha P. Tharp, and Amon Haruta, "Online Staffers: Superstars or Second-Class Citizens?" *Newspaper Research Journal* 20, no. 3 (1999): 29–47.

17. Jon E. Hyde, "Decoding the Codes: A Content Analysis of the News Coverage of Genetic Cloning by Three Online News Sites and Three National Daily Newspapers, 1996 through 1998" (Ph.D. diss., New York University, 2001).

18. Sharon Dunwoody, "Studying Users of the Why Files," *Science Communication* 22, no. 3 (2001): 274–82; William P. Eveland and Sharon Dunwoody, "Users and Navigation Patterns of a Science World Wide Web Site for the Public," *Public Understanding of Science* 7, no. 4 (1998): 285–311.

19. Peter Downs, "Science Break-through Gets No Post Coverage," *St. Louis Journalism Review* (February 2000): 1, 9.

20. Lynne M. Walters and Timothy N. Walters, "The Four Seasons: Patterns of Cyclical Success in Public Relations Placements," *Southwestern Mass Communication Journal* 10 (1994): 86–97.

21. "Paul Raeburn Offers a Wire-Service View of Science Writing," *ScienceWriters* (Fall 1994): 9–12.

22. See the Appendix.

23. Risa B. Burns et al., "Newspaper Reporting of the Medical Literature," *Journal of General Internal Medicine* 10 (1995): 19–24; Charlene A. Caburnay et al., "The News on Health Behavior: Coverage of Diet, Activity, and Tobacco in Local Newspapers," *Health Education and Behavior* 30, no. 6 (2003): 709–22.

24. Wallace Ravven, "Completeness of Science Reporting: A Comparison of Wire Services and Metropolitan Newspapers" (master's thesis, San Jose State University, 1979).

25. Jim Hartz and Rick Chappell, *Worlds Apart: How the Distance between Science and Journalism Threatens America's Future* (Nashville: First Amendment Center, 1998), 22.

26. E. E. Dennis and J. McCartney, "Science Journalists on Metropolitan Dailies," *Journal of Environmental Education* 10 (1979): 9–15; David H. Weaver and G. Cleveland Wilhoit, *The American Journalist: A Portrait of U.S. News People and Their Work,* 2d ed. (Bloomington: Indiana University Press, 1991); Karl B. Rubin, "Shaping Medical News: Factors Involved, Constraints Encountered and the Sources and Demographics of U.S. Medical Journalists" (master's thesis, Ohio State University, 1992); J. Sean McCleneghan, "The 1993 Newspaper Science Reporter: Contributing, Creative, and Responsible," *Social Science Journal* 31, no. 4 (1994): 467–77.

27. Melinda Voss, "Checking the Pulse: Midwestern Reporters' Opinions on Their Ability to Report Health Care News," *American Journal of Public Health* 92, no. 7 (2002): 1158–60; "Journalists Are More Likely to Be College Graduates," April 10, 2003, http://www.poynter.org/content/content_view.asp?id=28790 (accessed August 15, 2003).

28. Sharon Dunwoody, "The Challenge for Scholars of Popularized Science Communication: Explaining Ourselves," *Public Understanding of Science* 1 (1992): 11–14.

29. William D. Garvey, *Communication: The Essence of Science* (Oxford: Pergamon Press, 1979).

30. Julie M. Hurd, "Models of Scientific Communications Systems," in *From Print to Electronic: The Transformation of Scientific Communication,* edited by Susan Y. Crawford, Julie M. Hurd, and Ann C. Weller (Medford, N.J.: Information Today, 1996).

31. Hartz and Chappell, *Worlds Apart,* 103.

32. Garvey, *Communication,* 63.

33. Paul Ginsparg, "First Steps towards Electronic Research Communication," *Computers in Physics* 8, no. 4 (1994): 390–96; Ginsparg, "@xxx.lanl.gov," *Los Alamos Science* 22 (1994): 156–65; Gary Taubes, "Publication by Electronic Mail Takes Physics by Storm," *Science* 259 (1993): 1246–48; Ibironke Lawal, "Scholarly Communication: The Use and Non-use of E-print Archives for the Dissemination of Scientific Information," *Issues in Science and Technology Librarianship,* no. 36 (2002).

34. Ned Wright, "Previous What's New in Cosmologys," September 24, 2002, http://www.astro.ucla.edu/~wright/old_new_cosmo.html (accessed August 15, 2003). I was among those who missed the preprint: Vincent Kiernan, "Physicists Find That Neutrinos Have Mass after All, Casting Doubt on 'Standard Model,'" *Chronicle of Higher Education* (June 12, 1998): A18.

35. Monika K. Krzyzanowska, Melania Pintilie, and Ian F. Tannock, "Factors Associated with Failure to Publish Large Randomized Trials Presented at an Oncology Meeting," *Journal of the American Medical Association* 290, no. 4 (2003): 495–501; Aaron E. Carroll et al., "Does Presentation Format at the Pediatric Academic Societies' Annual Meeting Predict Subsequent Publication?" *Pediatrics* 112, no. 6 (2003): 1238–41; Lisa M. Schwartz, Steven Woloshin, and Lisa Baczek, "Media Coverage of

Scientific Meetings: Too Much, Too Soon?" *Journal of the American Medical Association* 287, no. 21 (2002): 2859–63.

36. Peter Conrad, "Uses of Expertise: Sources, Quotes, and Voice in the Reporting of Genetics in the News," *Public Understanding of Science* 8 (1999): 285–302.

37. Howard Kurtz, "Premature Publication," *Washington Post,* October 8, 1994, C1; Alison Bass, "Sex in '90s: A New Look," *Boston Globe,* October 7, 1994, 1.

38. Gillian Page, Robert Campbell, and A. J. Meadows, *Journal Publishing,* rev. ed. (Cambridge: Cambridge University Press, 1997), 8.

39. Ibid.

40. Ellen Ruppel Shell, "The Hippocratic Wars," *New York Times Magazine,* June 28, 1998, 34–38.

41. Page, Campbell, and Meadows, *Journal Publishing,* 164; Michael S. Wilkes, "The Public Dissemination of Medical Research: Problems and Solutions," *Journal of Health Communication* 2 (1997): 3–15; Lawrence K. Altman, "Promises of Miracles: News Releases Go Where Journals Fear to Tread," *New York Times,* January 10, 1995, C3; Page, Campbell, and Meadows, *Journal Publishing,* 7.

42. Comments by Laura Garwin and Monica Bradford to the District of Columbia Science Writers Association, April 15, 1996.

43. Page, Campbell, and Meadows, *Journal Publishing.*

44. The AAAS's annual Form 990 filed with the Internal Revenue Service is available at http://www.guidestar.org.

45. Erica Frank, "Authors' Criteria for Selecting Journals," *Journal of the American Medical Association* 272, no. 2 (1994): 163–64; Daniel J. DeBehnke, Jeffrey A. Kline, and Richard D. Shih, "Research Fundamentals: Choosing an Appropriate Journal, Manuscript Preparation, and Interactions with Editors," *Academic Emergency Medicine* 8, no. 8 (2001): 844–50; Lynn Dirk, "From Laboratory to Scientific Literature: The Life and Death of Biomedical Research Results," *Science Communication* 18 (1996): 3–28; Alma Swan and Sheridan Brown, *What Authors Want: The ALPSP Research Study on the Motivations and Concerns of Contributors to Learned Journals* (West Sussex, England: Association of Learned Professional Society Publishers, 1999).

46. "Clinical Trial Results: Exploring the Dissemination Process," 1991, http://consensus.nih.gov/ta/008/trials.doc (accessed August 7, 2003).

47. Catherine D. DeAngelis and Phil B. Fontanarosa, "To *JAMA* Peer Reviewers— Thank You," *Journal of the American Medical Association* 291, no. 6 (2004): 743; Edwin van Teijlingen and Vanora Hundley, "Getting Your Paper to the Right Journal: A Case Study of an Academic Paper," *Journal of Advanced Nursing* 37, no. 6 (2002): 506–11; Felix S. Chew and Annemarie Relyea-Chew, "How Research Becomes Knowledge in Radiology," *American Journal of Roentgenology* 150 (1988): 31–37.

48. Marlene Cimons, "Cover Calculations," *HHMI Bulletin* (June 2002): 10–13.

49. Conrad, "Uses of Expertise"; Shirley Ramsey, "A Benchmark Study of Elaboration and Sourcing in Science Stories for Eight American Newspapers," *Journalism and Mass Communication Quarterly* 76, no. 1 (1999): 87–98.

50. Einsiedel, "Framing Science and Technology"; Emma Weitkamp, "British Newspapers Privilege Health and Medicine Topics over Other Science News," *Public Relations Review* 29 (2003): 321–33; Massimiano Bucchi and Renato G. Mazzolini, "Big Science, Little News: Science Coverage in the Italian Daily Press, 1946–1997," *Public Understanding of Science* 12, no. 1 (2003): 7–24.

51. Tania M. Bubela and Timothy A. Caulfield, "Do the Print Media 'Hype' Genetic Research? A Comparison of Newspaper Stories and Peer-Reviewed Research Papers," *Canadian Medical Association Journal* 170, no. 9 (2004): 1399–407.

52. Lisa McGuire and Julia Ann Kelly, "Subject and Source Trends in the *Star Tribune*'s Coverage of the Medical Literature," *Minnesota Medicine* 86, no. 4 (2003): 33–38.

53. Christopher Dornan, "The 'Problem' of Science and the Media: A Few Seminal Texts in Their Context, 1956–1965," *Journal of Communication Inquiry* 12 (1988): 53–70.

54. American Chemical Society, "American Chemical Society Embargo Policy," http://www.eurekalert.org/jrnls/acs/index.php?page=embargo (accessed August 15, 2003).

55. André Picard, "Getting on Track: How Scientific Journals and Mainstream Journalists Could Do a Better Job of Communicating with the Public," *Canadian Medical Association Journal* 166, no. 9 (2002): 1153–54; National Association of Science Writers, "Communicating Science News," http://www.nasw.org/csn/sciwri.htm (accessed August 15, 2003).

56. Peter Gorner, "Science Reporting Checklist," 2002, http://www.facsnet.org/tools/sci_tech/gorner.php (accessed 5 August 2003); Hartz and Chappell, *Worlds Apart,* 104.

57. Robert B. McCall, "Science and the Press: Like Oil and Water?" *American Psychologist* 43, no. 2 (1988): 87–94.

58. Stephen Lock, *A Difficult Balance: Editorial Peer Review in Medicine* (Philadelphia: ISI Press, 1986); D. G. Altman, "Statistical Reviewing for Medical Journals," *Statistics in Medicine* 17, no. 23 (1998): 2661–74; S. N. Goodman, D. G. Altman, and S. L. George, "Statistical Reviewing Policies of Medical Journals: Caveat Lector?" *Journal of General Internal Medicine* 13, no. 11 (1998): 753–56; T. O. Jefferson et al., "Editorial Peer-Review for Improving the Quality of Reports of Biomedical Studies," *Cochrane Library,* no. 1 (2003).

59. D. F. Horrobin, "The Philosophical Basis of Peer Review and the Suppression of Innovation," *Journal of the American Medical Association* 263, no. 10 (1990): 1438–41; Frank J. Tipler, "Refereed Journals: Do They Insure Quality or Enforce Orthodoxy?" *Progress in Complexity, Information and Design* 2.1/2.2 (2003).

60. Janice Hopkins Tanne, "Speed the Science," *Columbia Journalism Review* (July–August 1999): 16.

61. Linda Rosa et al., "A Close Look at Therapeutic Touch," *Journal of the American Medical Association* 279, no. 13 (1998): 1005–10; American Medical Association,

"Science News Update, Week of April 1, 1998," http://www.ama-assn.org/sci-pubs/sci-news/1998/snr0401.htm (accessed August 15, 2003).

62. Gina Kolata, "A Child's Paper Poses a Medical Challenge," *New York Times,* April 1, 1998, A1; Terrence Monmaney and Louis Sahagun, "4th-Grader's Study Rebuts Touch Therapy," *Los Angeles Times,* April 1, 1998, A1; Brenda C. Coleman, "A Touchy Subject—Fourth-Grader Concludes Touch Therapy Is Quackery," Associated Press, April 1, 1998, Lexis-Nexis.

63. Mark Hagland, "Journals as Press Agents," *HMS Beagle* (August 7, 1998); Mary Ireland, "An Even Closer Look at Therapeutic Touch," *Journal of the American Medical Association* 280, no. 22 (1998): 1906; Susan M. Schmidt, "An Even Closer Look at Therapeutic Touch," *Journal of the American Medical Association* 280, no. 22 (1998): 1906; Andrew Freinkel, "An Even Closer Look at Therapeutic Touch," *Journal of the American Medical Association* 280, no. 22 (1998): 1905.

64. "Definition of 'Sole Contribution,'" *New England Journal of Medicine* 281 (1969): 676–77.

65. Stephen Lenier and Michael L.J. Apuzzo, "Response to 'The Public Dissemination of Medical Research: Problems and Solutions,'" *Journal of Health Communication* 2 (1997): 61–62.

66. Marcia Angell and Jerome P. Kassirer, "The Ingelfinger Rule Revisited," *New England Journal of Medicine* 325 (1991): 1371–73; Angell, "Multiple Sclerosis and the Ingelfinger Rule," *New England Journal of Medicine* 308 (1983): 217–18; Eliot Marshall, "Embargoes: Good, Bad or 'Necessary Evil'?" *Science* (October 30, 1998): 860–67; Phil B. Fontanarosa, Annette Flanagin, and Catherine D. DeAngelis, "The Journal's Policy Regarding Release of Information to the Public," *Journal of the American Medical Association* 284, no. 22 (2000): 2929–31.

67. Cary P. Gross et al., "Relation between Prepublication Release of Clinical Trial Results and the Practice of Carotid Endarterectomy," *Journal of the American Medical Association* 284, no. 22 (2000): 2886–93, Phil B. Fontanarosa and Annette Flanagin, "Prepublication Release of Medical Research," *Journal of the American Medical Association* 284, no. 22 (2000): 2927–29.

68. Elizabeth Gadd, Charles Oppenheim, and Steve Probets, "RoMEO Studies 4: An Analysis of Journal Publishers' Copyright Agreements," *Learned Publishing* 16, no. 4 (2003): 293–308; International Committee of Medical Journal Editors, "Uniform Requirements for Manuscripts Submitted to Biomedical Journals," *Journal of the American Medical Association* 277, no. 11 (1997): 927–34; American Association for the Advancement of Science, "What Is *Science*'s Embargo Policy? Can I Present Work Pending at *Science* at a Scientific Meeting?" http://www.sciencemag.org/feature/contribinfo/faq/embargo_faq.sht ml (accessed August 15, 2003); Michael S. Wilkes and Richard L. Kravitz, "Policies, Practices, and Attitudes of North American Medical Journal Editors," *Journal of General Internal Medicine* 10 (1995): 443–50.

69. Lawrence D. Grouse, "The Ingelfinger Rule," *Journal of the American Medical Association* 245 (1981): 375–76; Gail McBride, "Now for the Latest News," *Journal of the*

American Medical Association 245 (1981): 374–75; Lawrence K. Altman, "The Ingelfinger Rule, Embargoes, and Journal Peer Review—Part 1," *Lancet* 347 (1996): 1382–86; Nancy Ethiel, ed., *Medicine and the Media: A Changing Relationship,* Cantigny Conference Series (Chicago: Robert R. McCormick Tribune Foundation, 1995).

70. Lawrence K. Altman, "The Ingelfinger Rule, Embargoes, and Journal Peer Review—Part 2," *Lancet* 347 (1996): 1459–63.

71. American Association for the Advancement of Science, "What Is *Science*'s Embargo Policy?"; Philip Campbell, "*Nature* Embargo Policy," http://www.nature .com/nature/author/embargo.html (accessed August 15, 2003); "*Nature* Publication Policies," http://www.nature.com/nature/submit/policies/index.html (accessed January 6, 2006).

72. Eliot Marshall, "Scientific Meetings Produce Clash of Agendas," *Science* (October 30, 1998): 867–68.

73. Andrew S. Holtz, "When Should the Public Be Informed of the Results of Medical Research?" *Journal of the American Medical Association* 286, no. 23 (2001): 2944.

74. "PLoS Biology Editorial and Publishing Policies," (2004), http://biology.plos-journals.org/perlserve/?request=get-static&name=policies (accessed January 9, 2006).

75. Phil B. Fontanarosa and Catherine D. DeAngelis, "The Importance of the Journal Embargo," *Journal of the American Medical Association* 288, no. 6 (2002): 748–50; "Clinical Trial Results."

76. Jason Salzman, *Making the News: A Guide for Nonprofits and Activists* (Boulder, Colo.: Westview Press, 1998), 97; "Web Site Marks 5–Year Birthday," *Science* 292, no. 5521 (2001): 1564; "AAAS Science News Web Site Gaining Worldwide Audience," *Science* 300, no. 5619 (2003): 604–5.

77. "About the AlphaGalileo Project," http://www.alphagalileo.org/index.cfm?fuseaction=background (accessed January 22, 2005); Laura Miles, "AlphaGalileo: A Portal to Euro Science," *ScienceWriters* (Summer 2003): 7.

78. George D. Lundberg, "Providing Reliable Medical Information to the Public—Caveat Lector," *Journal of the American Medical Association* 262 (1989): 945; Fontanarosa, Flanagin, and DeAngelis, "Journal's Policy."

79. Arnold S. Relman, "Reporting the Aspirin Study: The Journal and the Media," *New England Journal of Medicine* 318 (1988): 918–20; Jerome P. Kassirer, "Dissemination of Medical Information: A Journal's Role," in *Doing More Good Than Harm: The Evaluation of Health Care Interventions,* edited by Kenneth S. Warren and Frederick Mosteller; *Annals of the New York Academy of Sciences* (New York: New York Academy of Sciences, 1993); Robert Steinbrook, "Medical Journals and Medical Reporting," *New England Journal of Medicine* 342 (2000): 1668–71.

80. Vikki Entwistle, "Reporting Research in Medical Journals and Newspapers," *BMJ* 310 (1995): 920–23.

81. Steven Woloshin and Lisa M. Schwartz, "Press Releases: Translating Research into News," *Journal of the American Medical Association* 287, no. 21 (2002): 2856–58;

Vincent Kiernan, "The Mars Meteorite: A Case Study in Controls on Dissemination of Science News," *Public Understanding of Science* 9 (2000): 15–41.

82. Rick Weiss, telephone interview by author, June 4, 2004; John Travis, telephone interview by author, June 4, 2004.

83. Mark Jurkowitz, "Why Embargoes May Aid Writers—and Readers," *Boston Globe,* December 9, 1999, E1, E6.

84. Floyd E. Bloom, "Embracing the Embargo," *Science,* 30 October 1998, 877. Embargoes also are no guarantee of equal treatment for all journalists. The AAAS considered giving the *New York Times* an additional embargo period, longer than other journalists', for coverage of the finding of possible fossilized bacteria on Mars. See Kiernan, "Mars Meteorite."

85. Timothy Johnson, "Medicine and the Media," *New England Journal of Medicine* 339 (1998): 87–92.

86. Comments by Peter Wrobel to the National Association of Science Writers, Denver, February 13, 2003; Weiss, telephone interview by author.

87. Carol H. Weiss, Eleanor Singer, and Phyllis M. Endreny, *Reporting of Social Science in the National Media* (New York: Russell Sage Foundation, 1988).

88. "White House Report Survives Release without an Embargo," *ScienceWriters* (Fall 1995): 3–5.

89. Harvey I. Leifert, "Who Broke the Embargo? (It's the Wrong Question)," *Physics Today* (October 2002): 48–49.

90. Tom Siegfried, "Reporting from Science Journals," in *A Field Guide for Science Writers, Second Edition* (New York: Oxford University Press, 2006), 13.

91. Ginger Pinholster, e-mail message to author, April 20, 2004.

92. Kiernan, "Mars Meteorite."

93. American Association for the Advancement of Science, "The Press Officer's Guide to *Science,*" http://www.eurekalert.org/pio/sci (accessed January 6, 2006).

94. Burns et al., "Newspaper Reporting"; Pinholster, e-mail message to author.

95. Kiernan, "Mars Meteorite"; Laurie P. Cohen and Antonio Regaldo, "How the Media Scurried to Tell Genome News," *Wall Street Journal,* February 13, 2001, B1, B4; Robert Lee Hotz, "Dolly, Dr. Seed, and Five Issues That Need Study," *ScienceWriters* (Winter 1997): 1–4.

96. Fontanarosa and DeAngelis, "Importance of the Embargo."

97. Patricia Anstett, "Hormone Therapy Is Risky," *Detroit Free Press,* July 9, 2002.

98. Fontanarosa and DeAngelis, "Importance of the Embargo."

99. "*JAMA* Claims Free Press Broke Embargo," http://www.poynter.org/forum/default.asp?=23219 (accessed June 7, 2004); Pat Anstett, telephone interview by author, April 28, 2004.

100. Mike King, "Stories on Major Medical News Can't Be Put on Hold," *Atlanta Journal-Constitution,* July 13, 2002, 9A; Michael Getler, "Hormones, Airlifts and Halliburton," *Washington Post,* July 14, 2002, B6; Kathryn S. Wenner, "News You Can't Use," *American Journalism Review* (September 2002): 12–13.

101. "*JAMA* Claims Free Press Broke Embargo"; Anstett, telephone interview by author; Carole Leigh Hutton to Catherine D. DeAngelis, July 9, 2002, http://www.poynter.org/forum/default.asp?id=23219 (accessed January 9, 2006).

102. Wenner, "News You Can't Use"; Richard Knox, "HRT Reaction," July 16, 2002, http://www.npr.org/rundowns/rundown.php?prgId=2&prgDate=16–Jul-2002 (accessed January 22, 2005); Anstett, telephone interview by author.

103. Fontanarosa and DeAngelis, "Importance of the Embargo."

104. James M. Bowler et al., "New Ages for Human Occupation and Climatic Change at Lake Mungo, Australia," *Nature* 412 (2003): 837–40; Bob Beale, "Embargoes and 'Nature' Journal," posting to ASC-L e-mail discussion list, February 25, 2003 (accessed April 29, 2004); David Appell, "Science Embargoes," February 26, 2003, http://nasw.org/users/appell/Weblog/indexold.html (accessed April 29, 2004); Bob Beale, e-mail message to author, May 19, 2004.

105. "Privacy Policy," http://www.eurekalert.org/privacy.php (accessed June 5, 2002).

106. Pamela J. Shoemaker and Stephen D. Reese, *Mediating the Message: Theories of Influences on Mass Media Content,* 2d ed. (White Plains, N.Y.: Longman, 1996), 125.

107. Timothy Crouse, *The Boys on the Bus* (New York: Ballantine, 1972); Robert Parry, *Fooling America: How Washington Insiders Twist the Truth and Manufacture the Conventional Wisdom,* 1st ed. (New York: William Morrow, 1992), 92.

108. Jane C. Lantz and William L. Lanier, "Observations from the Mayo Clinic National Conference on Medicine and the Media," *Mayo Clinic Proceedings* 77, no. 12 (2002): 1306–11.

109. Jay A. Winsten, "Science and the Media: The Boundaries of Truth," *Health Affairs* 4, no. 1 (1985): 5–23.

110. Karen Wright, "When Is a Breakthrough Really News?" *Los Angeles Times,* July 19, 1998, M1, M6.

111. Oscar Gandy, *Beyond Agenda-Setting: Information Subsidies and Public Policy* (Norwood, N.J.: Ablex, 1982), x.

112. Hillier Krieghbaum, *Science and the Mass Media* (New York: New York University Press, 1967).

113. Andrea H. Tanner, "Agenda Building, Source Selection, and Health News at Local Television Stations," *Science Communication* 25, no. 4 (2004): 350–63; David Murray, Joel Schwartz, and S. Robert Lichter, *It Ain't Necessarily So: How Media Make and Unmake the Scientific Picture of Reality* (Lanham, Md.: Rowman and Littlefield, 2001); John Abramson, "Medical Reporting in a Highly Commercialized Environment," *Nieman Reports,* Summer 2003, 54–57.

114. Andreas Frew, "Rumours and Whispers on the Way to the Moon," *New Scientist* (December 21–28, 1996): 79.

115. Christopher Bartlett, Jonathan Sterne, and Matthias Egger, "What Is Newsworthy? Longitudinal Study of the Reporting of Medical Research in Two British Newspapers," *BMJ* 325 (2002): 81–84; Entwistle, "Reporting Research."

116. Gaye Tuchman, *Making News: A Study in the Construction of Reality* (New York: Free Press, 1978); Michael Schudson, "Deadlines, Datelines and History," in *Reading the News,* edited by Robert Karl Manoff and Michael Schudson (New York: Pantheon Books, 1986), 81; Chester H. Rowell, "The Press as an Intermediary between the Investigator and the Public," *Science* 50, no. 1285 (1919): 146–50.

117. Charles Pekow, "Breaking the Pledge," *Quill* (November–December 1992): 23–25.

118. Denis McQuail, *Mass Communication Theory: An Introduction,* 3d ed. (London: Sage Publications, 1994).

119. J. Galtung and M. Ruge, "The Structure of Foreign News," *Journal of Peace Research* 1 (1965): 64–90; Shoemaker and Reese, *Mediating the Message.*

120. Boyce Rensberger, "Covering Science for Newspapers," in *A Field Guide for Science Writers,* edited by Deborah Blum and Mary Knudson (New York: Oxford University Press, 1997).

121. Burns et al., "Newspaper Reporting"; Jo Ellen Stryker, "Reporting Medical Information: Effects of Press Releases and Newsworthiness on Medical Journal Articles' Visibility in the News Media," *Preventive Medicine* 35 (2002): 519–30.

122. Richard C. Adelman and Lois M. Verbrugge, "Death Makes News: The Social Impact of Disease on Newspaper Coverage," *Journal of Health and Social Behavior* 41, no. 3 (2000): 347–67.

123. Vladimir de Semir, Christina Ribas, and Gemma Revuelta, "Press Releases of Science Journal Articles and Subsequent Newspaper Stories on the Same Topic," *Journal of the American Medical Association* 280 (1998): 294–95.

124. Cheryl Ann Abramson Thompson, "Science News Releases and News Coverage: Journalists' Opinions and Newspaper Usage" (master's thesis, Ohio State University, 1990).

125. Vincent Kiernan, "Embargoes and Science News," *Journalism and Mass Communication Quarterly* 80 (2004): 903–20.

126. Ibid.

Chapter 2: A Brief History of Embargoes in Science Journalism

1. Hillier Krieghbaum, "American Newspaper Reporting of Science News," *Kansas State College Bulletin* 25, no. 5 (1941); Richard L. Odiorne, "Science in the Newspapers since 1875" (bachelor's thesis, Massachusetts Institute of Technology, 1936); Normand Parent DuBeau, "Some Social Aspects of Science News" (master's thesis, University of Missouri, 1941); Marshall Missner, "Why Einstein Became Famous in America," *Social Studies of Science* 15, no. 2 (1985): 267–91. The *Times*'s story ran a day after coverage by British newspapers but before coverage by other American media. Missner notes that there are two competing explanations for the *Times*'s edge in breaking the story in the United States: One explanation is that the *Times*'s managing editor knew about the scientific conference at which the results were to be presented and

assigned a reporter to cover it. The second, which Missner finds more credible, is that the *Times*'s London reporter simply followed up on the coverage by the British newspapers.

2. J. A. Udden, "Science in Newspapers," *Popular Science Monthly* 84 (1914): 483–89.

3. Meyer Berger, *The Story of the "New York Times," 1851–1951* (New York: Simon and Schuster, 1951).

4. Edwin E. Slosson, "A New Agency for the Popularization of Science," *Science* 53 (1921): 321–23.

5. Edwin E. Slosson to John C. Merriam, February 21, 1921, Smithsonian Institution Archives, Records of Science Service, RU 7091, box 5, file "Carnegie Institution 1921"; Slosson to W. M. Gilbert, June 9, 1921, ibid.

6. Krieghbaum, "American Newspaper Reporting."

7. Ibid.

8. Carolyn D. Hay, "A History of Science Writers in the United States and of the National Association of Science Writers" (master's thesis, Northwestern University, 1970), 284–85.

9. Morris Fishbein, "Education in Chemistry and Medicine; Abstract of Address to the American Chemical Society," 1936, University of Chicago Library, Morris Fishbein Papers (hereafter Fishbein Papers), box 99, folder 6.

10. Robert D. Potter, "History of NASW," October 28, 1964, Archives of the National Association of Science Writers, Cornell University Library, accession no. 4448 (hereafter NASW Archives), box 4, folder 27.

11. James Stacey, "The Press Embargo: Friend or Foe?" *Journal of the American Medical Association* 254, no. 14 (1985): 1965–66; Morris Fishbein, *Morris Fishbein, M.D.: An Autobiography* (Garden City, N.Y.: Doubleday, 1969), 146, 211u.

12. Kent Cooper, *Kent Cooper and the Associated Press: An Autobiography* (New York: Random House, 1959), 111.

13. Arthur J. Snider to Executive Committee, National Association of Science Writers, June 4, 1964, Fishbein Papers, box 103, folder 6.

14. Stacey, "Press Embargo."

15. Nathaniel W. Faxon to Olin West, September 1, 1937, Fishbein Papers, box 99, folder 6; Morris Fishbein to N. W. Faxon, September 9, 1937, ibid.

16. Austin H. Clark, "The Work of the Press," *Science* 79 (1934): 141–42.

17. David Dietz, "Science and the American Press," *Science* 85 (1937): 107–12.

18. Krieghbaum, "American Newspaper Reporting"; William L. Laurence, "The Reminiscences of William L. Laurence," ed. interviews by Louis M. Starr and Scott Bruns (Columbia University oral history collection, microfiche, 1964), 220–21.

19. Sidney S. Negus, "For A. Ph. A. Section Chairmen," 1940, Austin H. Clark Papers, Smithsonian Institution, RU 7183, box 9, folder 5; emphasis in original.

20. Austin H. Clark, "Science Progress through Publicity," *Scientific Monthly* 52 (1941): 257–60.

21. Herbert B. Nichols, "Abstracts for the Press," *Scientific Monthly* 65 (1947): 405–7.

22. F. Barrows Colton, "Some of My Best Friends Are Scientists," *Scientific Monthly* 69 (1949): 156–60.

23. Sidney S. Negus, "For Secretaries and Program Chairmen of AAAS Sections and Affiliated and Associated Societies Meeting in New York City, December 26–31, 1949," 1949, Clark Papers, RU 7183, box 9, folder 5.

24. Marvin Howard Alisky, "A Study of the Sunday Science Reporting in the *New York Times* and the *New York Herald Tribune,* from September 8, 1946, to June 1, 1947" (master's thesis, University of Texas at Austin, 1947).

25. Mae Eanes, "A.M.A. Report," *Newsletter of the National Association of Science Writers,* September 1, 1956, NASW Archives, box 5, folder 17; Eanes, "Report from A.M.A.," *Newsletter of the National Association of Science Writers,* June 1, 1957, NASW Archives, box 5, folder 18; John Troan, "Diagnosing Medical News," paper presented at the eighth annual public relations conference of the Medical Society of the State of North Carolina, Raleigh, February 25, 1955, 4.

26. Sidney S. Negus, "Public Information Service," *Science* 127 (1958): 409–10.

27. H. P. Rusch, "Letter from H. P. Rush to Jane Stafford, November 27, 1950, Science Service Records, RU 7091, Smithsonian Institution Archives (hereafter Science Service Records), box 294, folder 14; Jotham Johnson to Watson Davis, June 7, 1950, ibid., folder 4.

28. Christopher Dornan, "The 'Problem' of Science and the Media: A Few Seminal Texts in Their Context, 1956–1965," *Journal of Communication Inquiry* 12 (1988): 53–70.

29. Arthur J. Snider, "A Science Writer Has His Problems, Including the Habits of Scientists," *Quill* (October 1955): 14–16.

30. Marc Selvaggio, "The Making of Jonas Salk," *Pittsburgh* (June 1984): 42–51; Val Adams, "Release Broken by N.B.C. on Polio," *New York Times,* April 13, 1955, 40; Jane S. Smith, *Patenting the Sun: Polio and the Salk Vaccine* (New York: William Morrow, 1990), 318–19.

31. Snider, "Science Writer Has His Problems"; Dael Wolfle to Paul Klopsteg et al. September 30, 1957, Archives of the American Association for the Advancement of Science (hereafter AAAS Archives), box R-22-2, file "Editorial Board and Board of Directors—1957: II."

32. Bill Barton, "Report from Los Angeles," *Newsletter of the National Association of Science Writers,* October 1, 1953, NASW Archives, box 5, folder 14.

33. Earl Ubell and Pierre Fraley, "Report of the Committee on Release Time," *Newsletter of the National Association of Science Writers* 2, no. 3 (1954), NASW Archives, box 5, folder 15.

34. Arthur J. Snider, "When Is News News?" *Newsletter of the National Association of Science Writers,* June 1, 1959, ibid., folder 20; "Editor Is Expelled," *New York Times,* April 1, 1959, 18.

35. S. A. Goudsmit, "Publicity," *Physical Review Letters* 4 (1960): 1–2.

36. S. A. Goudsmit, "Press Coverage," 1974, Samuel Goudsmit Papers, Niels Bohr Library, Center for History of Physics, American Institute of Physics, College Park, Md. (hereafter Goudsmit Papers), box 49, folder XI.33; Mary Paul Paye, "Newspaper Science Writing: Its Social Function and Practice by Five National Science Reporters" (Ph.D. diss., Syracuse University, 1965), 288.

37. David Warren Burkett, *Writing Science News for the Mass Media* (Houston: Gulf Publishing, 1965), 149–62.

38. J. D. Hunley, ed., *The Birth of NASA: The Diary of T. Keith Glennan* (Washington, D.C.: U.S. Government Printing Office, 1993); "Memorandum for the Administrator," December 14, 1960, archives of the National Aeronautics and Space Administration, file "Office of Public Affairs," box 1, folder "PAO Correspondence (1960s)"; T. Keith Glennan to Earl Ubell, December 23, 1960, ibid.

39. Robert C. Toth, "When Should the Taxpayer Be Told?" *Newsletter of the National Association of Science Writers,* March 1963, NASW Archives, box 5, folder 27; James L. Kauffman, *Selling Outer Space: Kennedy, the Media, and Funding for Project Apollo, 1961–63* (Tuscaloosa: University of Alabama Press, 1994); Bruce V. Lewenstein, "NASA and the Public Understanding of Space Science," *Journal of the British Interplanetary Society* 46 (1993): 251–54.

40. Burkett, *Writing Science News,* 127; Joseph H. Kuney, "The Role of Public Relations in Science News Reporting" (master's thesis, American University, 1962), 28; "Science—Report to the Board of Directors, 26 and 27 March 1966," 1966, AAAS Archives, box 2-1-3, file "Minutes & Agenda, Jan.–March 1966."

41. Victor Cohn, "A Science Writer's View," *AAAS Bulletin,* January 1962, 5.

42. John Foster, *Science Writer's Guide* (New York: Columbia University Press, 1963), 34.

43. Alma Louise Nahigian, "Medicine and the Mass Media: A Study of Medical and Mass Media Efforts to Improve the Dissemination of Accurate Medical Information to the Public" (master's thesis, Boston University, 1963), 83, 94n11.

44. John Lannan, "What Do the Scientists Do at All Those Conventions?" *Newsletter of the National Association of Science Writers,* March 1965, NASW Archives, box 5, folder 31.

45. Hillier Krieghbaum, *Science and the Mass Media* (New York: New York University Press, 1967), 100; Nate Haseltine, "Editor's Page," *Newsletter of the National Association of Science Writers,* March 1967, NASW Archives, box 5, folder 36.

46. National Association of Science Writers, *A Handbook for Press Arrangements at Scientific Meetings* (Port Washington, N.Y.: National Association of Science Writers, 1962), 8.

47. Robert C. Cowen, "President's Page," *Newsletter of the National Association of Science Writers,* March 1968, NASW Archives, box 5, folder 38; Dael Wolfle, "Memorandum of Discussion with Representatives of the National Association of Science Writers on 8 March 1968 Concerning Arrangements for Pressroom at AAAS Annual

Meeting," *Newsletter of the National Association of Science Writers,* September 1968, ibid., folder 39.

48. Watson Davis, "Hucksters in the Temple," *Newsletter of the National Association of Science Writers,* June 1963, ibid., folder 29.

49. Gary Brooten, "Gatlinburg 1966," *Newsletter of the National Association of Science Writers,* December 1966, ibid., folder 35.

50. "Definition of 'Sole Contribution,'" *New England Journal of Medicine* 281 (1969): 676–77.

51. Leonard S. Zahn, "Editor's Page," *Newsletter of the National Association of Science Writers,* December 1969, NASW Archives, box 5, folder 42; Franz J. Ingelfinger, "Purpose of the General Medical Journal," *New England Journal of Medicine* 287 (1972): 1043.

52. Barbara J. Culliton, "Dual Publication: 'Ingelfinger Rule' Debated by Scientists and Press," *Science* 176 (1972): 1403–5.

53. Ibid.; Arnold S. Relman, "Medical Meetings Should Be Backgrounders, Not News," *NASW Newsletter,* November 1979, 9–10; Mark Bloom, "Relman Stands Alone at Meeting with Reporters," ibid., 10–11; Relman, "An Open Letter to the News Media," *New England Journal of Medicine* 300 (1979): 554–55; Relman, "The Ingelfinger Rule," *New England Journal of Medicine* 305 (1981): 824–26; Karen A. Frenkel, "*NYT* May Use FOI against *NEJM* Rule," *Newsletter of the National Association of Science Writers,* October 1981, NASW Archives, box 6, folder 7; Marcia Angell and Jerome P. Kassirer, "The Ingelfinger Rule," *New England Journal of Medicine* 326 (1992): 958; David Perlman, "The Ingelfinger Rule," *New England Journal of Medicine* 326 (1992): 957.

54. Stephen Lock, "Wie Es Eigentlich Gewesen Ist: International Co-operation among Medical Journals," *Ugeskrift for Laeger* 152 (1990): 3770.

55. International Committee of Medical Journal Editors, "Uniform Requirements for Manuscripts Submitted to Biomedical Journals: Writing and Editing for Biomedical Publication" (2005), http://www.icmje.org (accessed January 13, 2006).

56. Relman, "Ingelfinger Rule"; Relman, "More on the Ingelfinger Rule," *New England Journal of Medicine* 381, no. 17 (1988): 1125–26; Relman, "Medical Research, Medical Journals and the Public Interest," *Society of Research Administrators Journal* 21, no. 2 (1989): 7; Nicholas Wade, "Medical Journal Draws Lancet on Rival," *Science* 211 (1981): 561.

57. Culliton, "Dual Publication."

58. "More about Prepublication," *Newsletter of the National Association of Science Writers,* June–October 1973, NASW Archives, box 6, folder 1.

59. "Science: Report to the Board of Directors, 13–14 March 1971," 1971, AAAS Archives, box R-2-5, file "Publications Committee Meeting, May 1971"; "Report to the Editorial Board of Science," 1976, ibid., box R-22-2, file "Editorial Board—1976."

60. Diane Starr Petryk, "A Content Analysis of Medical News in Four Metropolitan Dailies" (master's thesis, Michigan State University, 1979); Samuel J. Cordes, "A Content Analysis of Medical News Reporting in the Pittsburgh Press and the Pittsburgh Post-Gazette" (master's thesis, Point Park College, 1985).

61. Bill Stockton, "Letters," *Newsletter of the National Association of Science Writers,* March 1972, NASW Archives, box 5, folder 47.

62. Michael Riordan, *The Hunting of the Quark: A True Story of Modern Physics* (New York: Simon and Schuster, 1987); Bruce Lewenstein, Howard Lewis, and James Cornell, transcript of interview of Walter Sullivan, 1995, Department of Communication, Cornell University, Ithaca, N.Y.; Sullivan, "New and Surprising Type of Atomic Particle Found," *New York Times,* November 17, 1974, 1, 29.

63. Audrey Likely, "Draft—Talk at NASW 50th Gala," 1984, NASW Archives, box 1, folder 47.

64. J. A. Krumhansl, George L. Trigg, and Gene L. Wells, "Publicity and Anti-publicity," *Physical Review Letters* 35 (1975): 819; Edwin L. Goldhasser to C. S. Wu, October 7, 1975, Goudsmit Papers, box 49, folder XI.33.

65. Sullivan to Goudsmit, December 24, 1975, Goudsmit Papers, box 49, folder XI.33.

66. "Highlights of the Executive and Council Meetings, New York City, 6 and 7 November 1975," *Bulletin of the American Physical Society* 2d ser., 21 (1976): 138; J. A. Krumhansl, George L. Trigg, and Gene L. Wells, "An Editorial Experiment," *Physical Review Letters* 36 (1976): 997–98.

67. John Lear, *Recombinant DNA: The Untold Story* (New York: Crown Publishers, 1978); Howard J. Lewis, "Before the Public Debate about Recombinant DNA—the Role of the Science Press," *ScienceWriters* (February 1985): 9–12.

68. Lear, *Recombinant DNA,* 119.

69. "Scientists Study Risks of New Organisms," *Monterey Peninsula Herald,* February 24, 1975, 11.

70. Lewis, "Before the Debate."

71. National Association of Science Writers, *Hold for Release: Embargoed Science, Embattled System* (Hedgesville, W.Va.: National Association of Science Writers, 1999); Lewis, "Before the Debate."

72. "Scientist Questions Proposed Lab Rules," *Monterey Peninsula Herald,* February 25, 1975, 4.

73. Lewis, "Before the Debate."

74. George Alexander, "Scientists to Resume Genetic Research," *Los Angeles Times,* February 28, 1975, 25; Stuart Auerbach, "Genetic Research Ban Lifted," *Washington Post,* February 28, 1975, 1; Jerry E. Bishop, "Gene-Grafting Research to Resume Soon under Strict Laboratory-Safety Measures," *Wall Street Journal,* February 28, 1975, 14; Robert Cooke, "Scientists to Resume Risky Work on Genes," *Boston Globe,* February 28, 1975, 1; Judith Randal, "Risky Research on Gene Changes Gets Go-Ahead," *Washington Star-News,* February 28, 1975, A2; Victor K. McElheny, "World Biologists Tighten Rules on 'Genetic Engineering' Work," *New York Times,* February 28, 1975, 1.

75. Transcriptions of the three broadcasts contain no mention of the conference. See Vanderbilt University, Television News Archive, http://tvnews.vanderbilt.edu (accessed January 8, 2003).

76. Donald S. Fredrickson, *The Recombinant DNA Controversy* (Washington, D.C.: ASM Press, 2001), 26.

77. National Association of Science Writers, *Hold for Release.*

78. Edward Edelson, "The President's Letter," *Newsletter of the National Association of Science Writers,* December 1980, NASW Archives, box 6, folder 7.

79. Sharon Dunwoody, "News-Gathering Behaviors of Specialty Reporters: A Two-Level Comparison of Mass Media Decision Making," *Newspaper Research Journal* 1, no. 1 (1979): 29–41; Dunwoody, "The Science Writing Inner Club: A Communication Link between Science and the Lay Public," *Science, Technology and Human Values* 5 (1980): 14–22; Susan Holly Stocking, "Mass Media Visibility of Medical School Research: The Role of Public Information Initiatives, Scientists' Publishing Activity, and Institutional Prestige" (Ph.D. diss., Indiana University, 1983), 71.

80. Jerry Bishop, "The Dunwoody Report: She Nailed Us to the Wall," *Newsletter of the National Association of Science Writers,* January 1980, NASW Archives, box 6, folder 7.

81. Mark Fitzgerald, "AMA Blacklists *Miami Herald,*" *Editor and Publisher* (May 3, 1986): 11–12; Fitzgerald, "Giving the American Medical Association a Headache," *Editor and Publisher* (May 24, 1986): 18.

82. Charles H. Hennekens, "Behind the Timetable," *New York Times,* February 27, 1988, 30.

83. Ibid.; Fred Molitor, "Accuracy in Science News Reporting by Newspapers: The Case of Aspirin for the Prevention of Heart Attacks," *Health Communication* 5 (1993): 209–24.

84. Lawrence K. Altman, "Aspirin Report Illustrates the Control of *New England Journal* on Data Flow," *New York Times,* January 28, 1988, A1.

85. I. Herbert Scheinberg, "When a Medical News Embargo Caused Harm," *New York Times,* April 30, 1994, 22; Jerome P. Kassirer, "Aspirin Study News Embargo Harmed No One," *New York Times,* May 14, 1994, 20.

86. Constance Holden, "Reuters Release of Aspirin Study Is *NEJM* Headache," *ScienceWriters* (Spring 1988): 1–3; Arnold S. Relman, "Reporting the Aspirin Study: The Journal and the Media," *New England Journal of Medicine* 318 (1988): 918–20; Brian Murphy, "Magazine Imposes Sanctions on News Service," Associated Press, February 9, 1988, Lexis-Nexis.

87. Janny Scott, "Wire Service Challenge Could Change the Way of Medical Reporting," *Los Angeles Times,* February 23, 1988, pt. 2, p. 4; Alex Beam, "Reuters to Relman: Drop Dead," *Boston Globe,* November 25, 1988, 85.

88. Rob Stein, "*New England Journal,* Reuters Reach Partial Settlement," United Press International, March 20, 1989, Lexis-Nexis; Pat Guy, "Reuters Breaks Off Talks on Embargo," *USA Today,* March 20, 1989, B2; Susan Okie, "Smoking Crack Cocaine May Cause Strokes in Young People," *Washington Post,* September 12, 1990, A6.

89. "Embargoes on Science?" *Nature* 332 (1988): 191; Daniel S. Greenberg, "Annals

of Nonsense: Reuters vs. the Ingelfinger Rule," *Science and Government Report* (March 15, 1988): 7.

90. Arnold S. Relman, "Our Readers Vote for the News Embargo," *New England Journal of Medicine* 318 (1988): 1680.

91. Constance Holden, "Reuters to Defy Journal Embargo," *Science* (February 19, 1988): 862; Alex S. Jones, "News Agency Defiance May Change How Medical Findings Are Reported," *New York Times,* February 11, 1988, A16.

92. Jerrold K. Footlick, *Truth and Consequences: How Colleges and Universities Meet Public Crises* (Phoenix: American Council on Education and Oryx Press, 1997).

93. Jean L. Marx, "The CF Gene Hits the News," *Science* (September 1, 1989): 924.

94. Survey Working Group, "NASW Membership Survey Yields a Few Surprises," *ScienceWriters* (Summer 1990): 4–7.

95. Relman, "Medical Research, Journals and Public Interest."

96. Larry Thompson, "NIH Director Delays Human Gene Experiment," *Washington Post,* October 19, 1988, A3; Larry Thompson, "Plan to Implant Foreign Genes Passes Key Regulatory Hurdle," *Washington Post,* December 10, 1988, A2; Barbara J. Culliton, "Journals and Data Disclosure," *Science* (November 11, 1988): 857; Rick Weiss, "Subcommittee Okays Human Gene Transfer," *Science News* (December 17, 1988): 389.

97. "*JAMA* to Change Publication Date," *Chronicle of Higher Education* (February 28, 1990): A6.

98. Vincent Kiernan, "Changing Embargoes and the *New York Times*' Coverage of the *Journal of the American Medical Association*," *Science Communication* 19 (1998): 212–21.

99. Karl B. Rubin, "Shaping Medical News: Factors Involved, Constraints Encountered and the Sources and Demographics of U.S. Medical Journalists" (master's thesis, Ohio State University, 1992), 53.

100. U.S. Senate, Special Committee on Aging, *Breakthroughs in Brain Research: A National Strategy to Save Billions in Health Care Costs* (Washington, D.C.: U.S. Government Printing Office, 1996), 75; ABC News, "Protests Made against Cuts in Alzheimer's Research," transcript 5127-6, June 17, 1995, Lexis-Nexis.

101. "Next Week in *Nature,*" *Nature* (June 22, 1995): ix; "As Time Goes By . . . Another Quarrel about Embargoes," *ScienceWriters* (Summer 1995): 4–7.

102. Wyeth-Ayerst Laboratories, "Weight Loss Drug Update," PR Newswire, August 27, 1996, Lexis-Nexis; Laura Johannes, "How Should Medical News Be Distributed?" *Wall Street Journal,* August 29, 1996, B1, B6.

103. David Baron, "An Open Letter to the AAAS," *ScienceWriters* (Spring 1994): 5–6; Nan Broadbent, "Response from the AAAS," *ScienceWriters* (Spring 1994): 6–7.

104. Gina Kolata, *Clone: The Road to Dolly, and the Path Ahead* (New York: William Morrow, 1998), 32.

105. McKie, "Scientists Clone Adult Sheep," *Observer,* February 23, 1997, 1; Kolata, *Clone,* 32.

106. Thomas H. Maugh II, "Scientists Report Cloning Adult Mammal," *Los Angeles Times,* February 23, 1997, 1; Robert Lee Hotz, "Dolly, Dr. Seed, and Five Issues That Need Study," *ScienceWriters* (Winter 1997): 1–4.

107. Richard Gooding, "Top Clinton Aide and the Sexy Call Girl," *Star,* September 10, 1996, 6–8, 33.

108. Vincent Kiernan, "The Mars Meteorite: A Case Study in Controls on Dissemination of Science News," *Public Understanding of Science* 9 (2000): 15–41.

109. Laurie P. Cohen and Antonio Regaldo, "How the Media Scurried to Tell Genome News," *Wall Street Journal,* February 13, 2001, B1, B4; John Sulston and Georgina Ferry, *The Common Thread: A Story of Science, Politics, Ethics, and the Human Genome* (Washington, D.C.: Joseph Henry Press, 2002), 241–45.

110. Boyce Rensberger, "Covering Science for Newspapers," in *A Field Guide for Science Writers,* edited by Deborah Blum and Mary Knudson (New York: Oxford University Press, 1997), 10.

111. Lawrence K. Altman, "The Ingelfinger Rule, Embargoes, and Journal Peer Review—Part 1," *Lancet* 347 (1996): 1382–86; Altman, "The Ingelfinger Rule, Embargoes, and Journal Peer Review—Part 2," *Lancet* 347 (1996): 1459–63; Richard Horton, "Ruling Out Ingelfinger?" *Lancet* 347 (1996): 1423–24.

112. Michele Benjamin, "Editing a Medical Journal," *European Science Editing* (January 1992): 25–26.

113. "About Quadnet," http://www.quad-net.com/qnet_subscribe.html (accessed December 10, 1998); "About Newswise," http://www.newswise.com/aboutnewswise.htm (accessed December 10, 1998).

114. Eliot Marshall, "Embargoes: Good, Bad or 'Necessary Evil'?" *Science* (October 30, 1998): 860–67; National Association of Science Writers, *Hold for Release;* Howard J. Lewis, "Journal Embargoes: So Often Impeached, yet Never Convicted," *ScienceWriters* (Winter 1998–1999): 3–6.

115. Steve Maran, interview by author, September 26, 1996.

116. Carol Cruzan Morton, "In How Many Ways Do Science Writers Love the Internet?" *ScienceWriters* (Summer 1996): 21–26.

117. Jerome P. Kassirer and Marcia Angell, "The Internet and the Journal," *New England Journal of Medicine* 332 (1995): 1709–10; Ronald E. LaPorte et al., "The Death of Biomedical Journals," *BMJ* 310 (1995): 1387–90.

Chapter 3: Accuracy in Science Journalism

1. Francine Grodstein et al., "Postmenopausal Hormone Therapy and Mortality," *New England Journal of Medicine* 336, no. 25 (1997): 1769–75.

2. David Brown, "Women's Use of Hormones Has Benefits, Risks," *Washington Post,* June 19, 1997, A1, A16.

3. "Corrections," *Washington Post,* June 20, 1997, A2.

4. Charles H. Hennekens and Julie E. Buring, *Epidemiology in Medicine,* 1st ed. (Boston: Little, Brown, 1987).

5. Rita Rubin and Harrison L. Rogers, *Under the Microscope: The Relationship between Physicians and the News Media* (Nashville: Freedom Forum First Amendment Center at Vanderbilt University, 1993); Floyd E. Bloom, "Embracing the Embargo," *Science* (October 30, 1998): 877; Phil B. Fontanarosa and Catherine D. DeAngelis, "The Importance of the Journal Embargo," *Journal of the American Medical Association* 288, no. 6 (2002): 748–50.

6. Sissela Bok, *Lying: Moral Choice in Public and Private Life,* 2d Vintage Books ed. (New York: Vintage Books, 1999); John C. Merrill, *Journalism Ethics: Philosophical Foundations for News Media* (New York: St. Martin's Press, 1997); Society of Professional Journalists, "Code of Ethics," http://www.spj.org/ethics_code.asp (accessed August 18, 2003); American Society of Newspaper Editors, "ASNE Statement of Principles," August 28, 2002, http://www.asne.org/kiosk/archive/principl.htm (accessed January 9, 2006).

7. Commission on Freedom of the Press, *A Free and Responsible Press* (Chicago: University of Chicago Press, 1947), 21; James Fallows, *Breaking the News: How the Media Undermine American Democracy* (New York: Vintage Books, 1996); Leonard Downie and Robert G. Kaiser, *The News about the News: American Journalism in Peril,* 1st ed. (New York: Alfred A. Knopf, 2002).

8. Margaret E. Brunt et al., "Mass Media Release of Medical Research Results," *Journal of General Internal Medicine* 18, no. 2 (2003): 84–94. However, results of another similar study were more ambiguous: Malcolm Maclure et al., "Influences of Educational Interventions and Adverse News about Calcium-Channel Blockers on First-Line Prescribing of Antihypertensive Drugs to Elderly People in British Columbia," *Lancet* 352, no. 9132 (1998): 943–48.

9. Bloom, "Embracing the Embargo."

10. Scott Maier, "Getting It Right? Not in 59 Percent of Stories," *Newspaper Research Journal* 23, no. 1 (2002): 10–24.

11. Boyce Rensberger, "Covering Science for Newspapers," in *A Field Guide for Science Writers,* edited by Deborah Blum and Mary Knudson (New York: Oxford University Press, 1997), 16; Alicia C. Shepard, "Show and Print," *American Journalism Review* (March 1996): 40–44.

12. Sheryl Fragin, "Flawed Science at the *Times,*" *Brill's Content* (October 1998): 104–15; Mark Dowie, "What's Wrong with the *New York Times*'s Science Reporting?" *Nation* (July 8, 1998): 13–19; Gina Kolata, "A Cautious Awe Greets Drugs That Eradicate Tumors in Mice," *New York Times,* May 3, 1998, 1; Abigail Pogrebin, "The Mouse That Roared," *Brill's Content* (October 1998): 110–11; David Shaw, "Overdose of Optimism," *Los Angeles Times,* February 13, 2000, 1; Howard J. Lewis, "The Kolata Story: When No News Made Big News Over and Over," *ScienceWriters* (Spring–Summer 1998): 1; Eliot Marshall, "The Power of the Front Page of the *New York Times,*"

Science (May 15, 1998): 996–97; James D. Watson, "High Hopes on Cancer," *New York Times,* May 7, 1998, A30. I was among those who produced stories on antiangiogenesis in the ensuing days. See Vincent Kiernan, "Scientists Raise Doubts about Widely Publicized Cancer Drugs," *Chronicle of Higher Education* (May 15, 1998): A20.

13. Jane E. Brody, "The Challenges of Accurate Health Reporting," December 1, 2000, http://www.ahcj.umn.edu/files/brodyspeech.pdf (accessed July 31, 2003).

14. Cystale Purvis Cooper and Darcie Yukimura, "Science Writers' Reactions to a Medical 'Breakthrough' Story," *Social Science and Medicine* 54, no. 12 (2002): 1887–96.

15. Harold D. Lasswell, "The Structure and Function of Communication in Society," in *The Communication of Ideas,* edited by Lyman Bryson (New York: Harper and Brothers, 1948), 37.

16. Claude Elwood Shannon and Warren Weaver, *The Mathematical Theory of Communication* (Urbana: University of Illinois Press, 1949).

17. Kenneth J. Gergen, *An Invitation to Social Construction* (London: Sage, 1999), 36–37.

18. Sharon Dunwoody, "Community Structure and Media Risk Coverage," *RISK: Health, Safety and Environment* 5 (1994): 193–201.

19. John E. Newhagen and Mark R. Levy, "The Future of Journalism in a Distributed Communication Architecture," in *The Electronic Grapevine: Rumor, Reputation, and Reporting in the New On-line Environment,* edited by Diane L. Borden and Kerric Harvey (Mahwah, N.J.: Lawrence Erlbaum Associates, 1998), 13.

20. Jay A. Winsten, "Science and the Media: The Boundaries of Truth," *Health Affairs* 4, no. 1 (1985): 5–23; Dorothy Nelkin, *Selling Science: How the Press Covers Science and Technology,* rev. ed. (New York: W. H. Freeman, 1995); E. W. Campion, "Medical Research and the News Media," *New England Journal of Medicine* 351, no. 23 (2004): 2436–37.

21. Sharon Dunwoody, "Scientists, Journalists, and the Meaning of Uncertainty," in *Communicating Uncertainty: Media Coverage of New and Controversial Science,* edited by Sharon M. Friedman, Sharon Dunwoody, and Carol L. Rogers (Mahwah, N.J.: Lawrence Erlbaum Associates, 1999), 76.

22. Mitchell V. Charnley, "Preliminary Notes on a Study of Newspaper Accuracy," *Journalism Quarterly* 13, no. 2 (1936): 394–400.

23. Charles W. Finley and Otis W. Caldwell, *Biology in the Public Press* (New York: Lincoln School of Teachers College, 1923), 144–45.

24. James W. Tankard Jr. and Michael Ryan, "News Source Perceptions of Accuracy of Science Coverage," *Journalism Quarterly* 51 (1974): 219–25, 334; Barbara Moore and Michael Singletary, "Scientific Sources' Perceptions of Network News Accuracy," *Journalism Quarterly* 62 (1985): 816–23.

25. Beth Heffner, "Communicatory Accuracy: Four Experiments," *Journalism Monographs* 30 (1973); Michael Ryan, "A Factor Analytic Study of Scientists' Responses to Errors," *Journalism Quarterly* 52, no. 2 (1975): 333–36.

26. D. Lynn Pulford, "Follow-up Study of Science News Accuracy," *Journalism Quarterly* 53 (1976): 119–21.

27. Celeste Condit, "Science Reporting to the Public: Does the Message Get Twisted?" *Canadian Medical Association Journal* 170, no. 9 (2004): 1415–16; Robert B. McCall and S. Holly Stocking, "Between Scientists and Public: Communicating Psychological Research through the Mass Media," *American Psychologist* 37 (1982): 985–95; Sharon Dunwoody, "A Question of Accuracy," *IEEE Transactions on Professional Communication* 25, no. 4 (1982): 196–99.

28. Laura D. Carsten and Deborah L. Illman, "Perceptions of Accuracy in Science Writing," *IEEE Transactions on Professional Communication* 45, no. 3 (2002): 153–56.

29. Peter Guy Northouse and Laurel Lindhout Northouse, *Health Communication: Strategies for Health Professionals*, 2d ed. (Norwalk, Conn.: Appleton and Lange, 1992).

30. Eleanor Singer, "A Question of Accuracy: How Journalists and Scientists Report Research on Hazards," *Journal of Communication* 40, no. 4 (1990): 102–16; Anne Moyer et al., "Accuracy of Health Research Reported in the Popular Press: Breast Cancer and Mammography," *Health Communication* 7, no. 2 (1995): 147–61; Jane Wells et al., "Newspaper Reporting of Screening Mammography," *Annals of Internal Medicine* 135, no. 12 (2001): 1029–37; Singer, "A Question of Accuracy"; Raymond N. Ankney, Patricia Heilman, and Jacob Kolff, "Newspaper Coverage of the Coronary Artery Bypass Grafting Report," *Science Communication* 18 (1996): 153–64; Evette M. Hackman and Gaile L. Moe, "Evaluation of Newspaper Reports of Nutrition-Related Research," *Journal of the American Dietetic Association* 99, no. 12 (1999): 1564–66; Megan M. MacDonald and Laurie Hoffman-Goetz, "A Retrospective Study of the Accuracy of Cancer Information in Ontario Daily Newspapers," *Canadian Journal of Public Health* 93, no. 2 (2002): 142–45.

31. Tania M. Bubela and Timothy A. Caulfield, "Do the Print Media 'Hype' Genetic Research? A Comparison of Newspaper Stories and Peer-Reviewed Research Papers," *Canadian Medical Association Journal* 170, no. 9 (2004): 1399–407; Condit, "Science Reporting to the Public."

32. Andrew D. Oxman et al., "An Index of Scientific Quality for Health Reports in the Lay Press," *Journal of Clinical Epidemiology* 46, no. 9 (1993): 987–1001; Frank J. Molnar et al., "Assessing the Quality of Newspaper Medical Advice Columns for Elderly Readers," *Canadian Medical Association Journal* 161, no. 4 (1999): 393–95; Eliza Mountcastle-Shah et al., "Assessing Mass Media Reporting of Disease-Related Genetic Discoveries: Development of an Instrument and Initial Findings," *Science Communication* 24, no. 4 (2003): 458–78.

33. David Murray, Joel Schwartz, and S. Robert Lichter, *It Ain't Necessarily So: How Media Make and Unmake the Scientific Picture of Reality* (Lanham, Md.: Rowman and Littlefield, 2001), 188.

34. William A. Evans et al., "Science in the Prestige and National Tabloid Presses,"

Social Science Quarterly 71 (1990): 105–17; Marianne G. Pellechia, "Trends in Science Coverage: A Content Analysis of Three U.S. Newspapers," *Public Understanding of Science* 6 (1997): 49–68.

35. Melinda Voss, "Why Reporters and Editors Get Health Coverage Wrong," *Nieman Reports* (Spring 2003): 46–48.

36. Ray Moynihan et al., "Coverage by the News Media of the Benefits and Risks of Medications," *New England Journal of Medicine* 342, no. 22 (2000): 1645–50; Alan Cassels et al., "Drugs in the News: An Analysis of Canadian Newspaper Coverage of New Prescription Drugs," *Canadian Medical Association Journal* 168, no. 9 (2003): 1133–37.

37. Lisa M. Schwartz and Steven Woloshin, "The Media Matter: A Call for Straightforward Medical Reporting," *Annals of Internal Medicine* 140, no. 3 (2004): 226–28; John Abramson, "Medical Reporting in a Highly Commercialized Environment," *Nieman Reports* (Summer 2003): 54–57; Moynihan et al., "Coverage of Benefits and Risks."

38. Warren Weaver, "Communicative Accuracy," *Science* 127, no. 7 March (1958): 499.

39. Phillip J. Tichenor et al., "Mass Communication Systems and Communication Accuracy in Science News Reporting," *Journalism Quarterly* 47 (1970): 673–83.

40. Edward J. Lordan, "Do Methodological Details Help Readers Evaluate Statistics-Based Stories?" *Newspaper Research Journal* 14, no. 3–4 (1993): 13–19.

41. Fred Molitor, "Accuracy in Science News Reporting by Newspapers: The Case of Aspirin for the Prevention of Heart Attacks," *Health Communication* 5 (1993): 209–24.

42. Susan Dentzer, "Media Mistakes in Coverage of the Institute of Medicine's Error Report," *Effective Clinical Practice* 3, no. 6 (2000): 305–8.

43. Vincent Kiernan, "The Mars Meteorite: A Case Study in Controls on Dissemination of Science News," *Public Understanding of Science* 9 (2000): 15–41.

44. J. William Schopf, *Cradle of Life: The Discovery of Earth's Earliest Fossils* (Princeton, N.J.: Princeton University Press, 1999); Charles Seife, "Money for Old Rock," *New Scientist*, 8 August 1998, 20–21; R. M. Holliman, "British Public Affairs Media and the Coverage of 'Life on Mars'?" in *Communicating Science: Contexts and Channels,* edited by Eileen Scanlon, Elizabeth Whitelegg, and Simeon Yates (London: Routledge, 1999).

45. Kiernan, "Mars Meteorite."

46. John Dudley Miller, "'NEJM' Reveals Error in Article on Bias; Reporters Erred, Too," *ScienceWriters* (Summer 1999): 4–5.

47. Gilbert L. Fowler and Tommy L. Mumert, "A Survey of Correction Policies of Arkansas Newspapers," *Journalism Quarterly* 65, no. 4 (1988): 853–58; Steve M. Barkin and Mark R. Levy, "All the News That's Fit to Correct: Corrections in the *Times* and the *Post*," *Journalism Quarterly* 60, no. 2 (1983): 218–25; Michael E. Cremedas, "Corrections Policies in Local Television News: A Survey," *Journalism Quarterly* 69, no. 1 (1992): 166–72.

48. Anna Larsson et al., "Medical Messages in the Media—Barriers and Solutions to Improving Medical Journalism," *Health Expectations* 6, no. 4 (2003): 323–31.

49. Donald P. Hayes, "The Growing Inaccessibility of Science," *Nature* 356 (1992): 739–40.

50. Sarah Cohen and Len Bruzzese, *Numbers in the Newsroom: Using Math and Statistics in News* (Columbia, Mo.: Investigative Reporters and Editors, 2001); Kathleen Wickham, *Math Tools for Journalists,* 2d ed. (Oak Park, Ill.: Marion Street Press, 2003); Melinda Voss, "Checking the Pulse: Midwestern Reporters' Opinions on Their Ability to Report Health Care News," *American Journal of Public Health* 92, no. 7 (2002): 1158–60.

51. J. Frank Yates, *Judgment and Decision Making* (Englewood Cliffs, N.J.: Prentice-Hall, 1990).

52. Larsson et al., "Medical Messages"; Voss, "Checking the Pulse"; Nelkin, *Selling Science.*

53. Kenneth R. Hammond, *Judgments under Stress* (New York: Oxford University Press, 2000).

54. S. Holly Stocking and Paget H. Gross, *How Do Journalists Think? A Proposal for the Study of Cognitive Bias in Newsmaking* (Bloomington, Ind.: ERIC Clearinghouse on Reading and Communication Skills, 1989), 71.

55. Robert Steinbrook, "Medical Journals and Medical Reporting," *New England Journal of Medicine* 342 (2000): 1668–71.

Chapter 4: Costs of the Embargo

1. Floyd E. Bloom, "Embracing the Embargo," *Science* (October 30, 1998): 877.

2. Harvey I. Leifert, "Who Broke the Embargo? (It's the Wrong Question)," *Physics Today* (October 2002): 48–49.

3. See the Appendix.

4. Jon Turney, "How to Commune with Nature," *Times Higher Education Supplement* (February 27, 1998): 17.

5. Anke M. van Trigt et al., "Journalists and Their Sources of Ideas and Information on Medicines," *Social Science and Medicine* 38 (1994): 637–43.

6. Pierre-Marie Fayard, "Science News in the European Dailies: The Invisible Network of the Agenda Setting," *Accountability in Research* 5 (1997): 140.

7. Massimiano Bucchi and Renato G. Mazzolini, "Big Science, Little News: Science Coverage in the Italian Daily Press, 1946–1997," *Public Understanding of Science* 12, no. 1 (2003): 7–24.

8. Marcelo Leite, "Reporting on Science in South America," *Nieman Reports* (Fall 2002): 42–44.

9. Miriam Shuchman and Michael S. Wilkes, "Medical Scientists and Health News Reporting: A Case of Miscommunication," *Annals of Internal Medicine* 126 (1997): 976–82.

10. Commission on Freedom of the Press, *A Free and Responsible Press* (Chicago: University of Chicago Press, 1947); Kathryn S. Wenner, "News You Can't Use," *American Journalism Review* (September 2002): 12–13.

11. Peter Cram et al., "The Impact of a Celebrity Promotional Campaign on the Use of Colon Cancer Screening: The Katie Couric Effect," *Archives of Internal Medicine* 163, no. 13 (2003): 1601–5; Peter Conrad, "Uses of Expertise: Sources, Quotes, and Voice in the Reporting of Genetics in the News," *Public Understanding of Science* 8 (1999): 285–302; Keay Davidson, "Superbomb Ignites Science Dispute," *San Francisco Chronicle,* September 28, 2003, A1; Sharon Weinberger, "Scary Things Come in Small Packages," *Washington Post Magazine,* March 28, 2004, W15; Karen Jegalian, "Flash in the Lab," *Dallas Morning News,* August 24, 1998, 8D.

12. Jocalyn P. Clark, Georgina D. Feldberg, and Paula A. Rochon, "Representation of Women's Health in General Medical versus Women's Health Specialty Journals: A Content Analysis," *BMC Women's Health* 2, no. 5 (2002).

13. Bloom, "Embracing the Embargo," 877.

14. Peter Gorner, "The Edgy State of Science," *Chicago Tribune,* February 13, 2000, sec. 2, pp. 1, 8, 9.

15. Matthew C. Nisbet et al., "Knowledge, Reservations, or Promise? A Media Effects Model for Public Perceptions of Science and Technology," *Communication Research* 29 (2002): 584–608.

16. Ruth E. Patterson et al., "Is There a Consumer Backlash against the Diet and Health Message?" *Journal of the American Dietetic Association* 101, no. 1 (2001): 37–41.

17. Gideon Koren and Naomi Klein, "Bias against Negative Studies in Newspaper Reports of Medical Research," *Journal of the American Medical Association* 266 (1991): 1824–26; Conrad, "Uses of Expertise."

18. Marcia Angell, *Science on Trial: The Clash of Medical Evidence and the Law in the Breast Implant Case* (New York: Norton, 1996), 171.

19. Marcia Angell and Jerome P. Kassirer, "Clinical Research—What Should the Public Believe?" *New England Journal of Medicine* 331 (1994): 189–90.

20. Nancy Ethiel, ed., *Medicine and the Media: A Changing Relationship,* Cantigny Conference Series (Chicago: Robert R. McCormick Tribune Foundation, 1995), 16–17.

21. Shannon Brownlee, "Health, Hope and Hype," *Washington Post,* August 3, 2003, B1, B5.

22. Everett M. Rogers, *Diffusion of Innovations,* 5th ed. (New York: Free Press, 2003), 169.

23. Bruce V. Lewenstein, "From Fax to Facts: Communication in the Cold Fusion Saga," *Social Studies of Science* 25 (1995): 403–36.

24. J. Willems and E. Woudstra, "The Use by Biologists and Engineers of Nonspecialist Information Sources and Its Relation to Their Social Involvement," *Scientometrics* 28 (1993): 205–16.

25. M. Timothy O'Keefe, "The Mass Media as Sources of Medical Information for Doctors," *Journalism Quarterly* 47 (1970): 95–100; Rita Rubin and Harrison L. Rogers, *Under the Microscope: The Relationship between Physicians and the News Media* (Nashville: Freedom Forum First Amendment Center at Vanderbilt University, 1993); Erica Frank, Grant Baldwin, and Alan M. Langlieb, "Continuing Medical Education Habits of U.S. Women Physicians," *Journal of the American Medical Women's Association* 55, no. 1 (2000): 27–28.

26. David P. Phillips et al., "Importance of the Lay Press in the Transmission of Medical Knowledge to the Scientific Community," *New England Journal of Medicine* 325 (1991): 1180–83.

27. Comments by Monica Bradford, managing editor of *Science,* to the District of Columbia Science Writers Association, April 15, 1996.

28. "The Impact Factor," http://sunweb.isinet.com/isi/hot/essays/journalcitation-reports/7. html (accessed August 5, 2003); Barbara Gastel, "Assessing the Impact of Investigators' Work: Beyond Impact Factors," *Canadian Journal of Anaesthesiology* 48, no. 10 (2001): 941–45; Phil B. Fontanarosa, "Impact Factors and Emergency Medicine Journals," *Annals of Emergency Medicine* 31, no. 1 (1998): 107–9; E. John Gallagher and Douglas P. Barnaby, "Evidence of Methodologic Bias in the Derivation of the Science Citation Index Impact Factor," *Annals of Emergency Medicine* 31, no. 1 (1998): 83–86; Douglas P. Barnaby and E. John Gallagher, "Alternative to the Science Citation Index Impact Factor as an Assessment of Emergency Medicine's Scientific Contributions," *Annals of Emergency Medicine* 31, no. 1 (1998): 78–82; Vasilis Theoharakis and Mary Skordia, "How Do Statisticians Perceive Statistics Journals?" *American Statistician* 57, no. 2 (2003): 115–23.

29. Somnath Saha, Sanjay Saint, and Dimitri A. Christakis, "Impact Factor: A Valid Measure of Journal Quality?" *Journal of the Medical Library Association* 91, no. 1 (2003): 42–46; Peter A. Lawrence, "The Politics of Publication," *Nature* 422 (2003): 259–61; Argyro Fassoulaki et al., "Academic Anesthesiologists' Views on the Importance of the Impact Factor of Scientific Journals: A North American and European Survey," *Canadian Journal of Anesthesiology* 48, no. 10 (2001): 953–57.

30. Robert L. Perlman, "Ethical Issues in Biomedical Publishing," *Perspectives in Biology and Medicine* 45, no. 1 (2002): 125–30.

31. Vincent Kiernan, "Diffusion of News about Research," *Science Communication* 25, no. 1 (2003): 3–13.

32. Laurel Richardson Walum, "Sociology and the Mass Media: Some Major Problems and Modest Proposals," *American Sociologist* 10, no. 2 (February 1975): 28–32.

33. Richard C. Horton, "Journals versus Journalists," *European Science Editing,* no. 54 (1995): 3–7.

34. Steven Woloshin and Lisa M. Schwartz, "Press Releases: Translating Research into News," *Journal of the American Medical Association* 287, no. 21 (2002): 2856–58.

35. Laura Johannes, "How Should Medical News Be Distributed?" *Wall Street Journal,* August 29, 1996, B1, B6.

36. *United States v. O'Hagan,* 521 U.S. 642 (1997).

37. J. R. Ferguson, "Biomedical Research and Insider Trading," *New England Journal of Medicine* 337, no. 9 (1997): 631–64.

38. The journalists may even show an embargoed article to the author's scientific competitors. See Michael Waldholz, *Curing Cancer: The Story of the Men and Women Unlocking the Secrets of Our Deadliest Illness* (New York: Simon and Schuster, 1997), 205.

39. Jerry E. Bishop, "Wall Street Insiders Cash in While Writers Observe *NEJM* Release Time," *Newsletter of the National Association of Science Writers,* June 1981, Archives of the National Association of Science Writers, Cornell University Library, accession no. 4448, box 6, folder 8; Bishop, "*NEJM* Cracks Down on Release Time—at Last," *Newsletter of the National Association of Science Writers,* April 1982, ibid., folder 7; Michael Waldholz, "E.T. They're Not, but 'Alien' Reporters Are Beating Science Writers on Key Stories," *Newsletter of the National Association of Science Writers,* August 1982, ibid., folder 8.

40. Wade Roush, "'Fat Hormone' Poses Hefty Problem for Journal Embargo," *Science* (August 4, 1995): 627; Paul Jacobs, "Profiting on Science Discoveries," *Los Angeles Times,* January 8, 1999, A1, A6; Scott Hensley, Peter A. McKay, and David P. Hamilton, "Leaky Lid on Science News," *Wall Street Journal,* July 25, 2002, B1, B4.

41. Geeta Anand and Randall Smith, "Biotech Analysts Strive to Peek Inside Clinical Tests of Drugs," *Wall Street Journal,* August 8, 2002, 1.

42. Sue Pelletier, "SEC Takes Stock of Medical Meetings," *Medical Meetings* (December 2002): 8.

43. Tom Abate, "Cancer Therapies at Border Where Medicine, Stock Market Collide," *San Francisco Chronicle,* April 22, 2002, E1; Naomi Aoki, "When Science and Finance Collide," *Boston Globe,* April 24, 2002, F1; Scott Gottlieb, "Research Should Be Released More Quickly," *Los Angeles Times,* June 17, 2001, 3; Wenner, "News You Can't Use"; Catherine Arnst, "How Drug News Leaks to Investors," *Business Week,* May 6, 2002, 32.

44. Jacobs, "Profiting on Science Discoveries."

45. Nancy Ethiel, ed., *Ethical Issues in the Publication of Medical Information,* Cantigny Conference Series (Chicago: Robert R. McCormick Tribune Foundation, 1999), 120; Andrew A. Skolnick, "SEC Going after Insider Trading Based on Medical Research Results," *Journal of the American Medical Association* (July 1, 1998): 10–11.

Chapter 5: The Embargo Should Go

1. Pia Pini, "Media Wars," *Lancet* 346 (1995): 1681.

2. "Journalists Speak Out for Open Access," *Open Access Now* (January 19, 2004): A2–A3; Robin McKie, telephone interview by author, May 10, 2004.

3. The largely moribund concept of a news council might serve as the model for a body to investigate journal embargo complaints. In 1992, the Minnesota News

Council ruled that WCCO-TV in Minneapolis violated an embargo on a report on sites for a new airport. The news council had no authority to punish the television station, but the ruling itself may have embarrassed the broadcaster. In the context of journal embargoes, allegations of embargo violations could be adjudicated by a panel of journalists and journal officials not connected with the incident in question. See Minnesota News Council, "Determination 92," http://www.mtn.org/~newscncl/complaints/hearings/det_92.html (accessed June 6, 2004).

4. Christopher Dornan, "Some Problems in Conceptualizing the Issue of 'Science and the Media,'" *Critical Studies in Mass Communication* 7 (1990): 48–71.

5. Bill Kovach and Tom Rosenstiel, *The Elements of Journalism: What Newspeople Should Know and the Public Should Expect,* 1st ed. (New York: Crown Publishers, 2001), 17.

6. Lisa M. Schwartz and Steven Woloshin, "On the Prevention and Treatment of Exaggeration," *Journal of General Internal Medicine* 18, no. 2 (2003): 153–54.

7. Caryl Rivers, "Good Reporters Make Best Science Writers," *Editor and Publisher* (January 23, 1965): 17, 52.

8. Grace Baynes, telephone interview by author, May 4, 2004.

9. Ibid.

10. Deborah Blum, "Investigating Science," *Nieman Reports* (Fall 2002): 13–15.

11. Robert Lee Hotz, "The Difficulty of Finding Impartial Sources in Science," *Nieman Reports* (Fall 2002): 6–7; National Association of Science Writers, *Hold for Release: Embargoed Science, Embattled System* (Hedgesville, W.Va.: National Association of Science Writers, 1999).

12. Dorothy Nelkin, *Selling Science: How the Press Covers Science and Technology,* rev. ed. (New York: W. H. Freeman, 1995), 100; Hotz, "Finding Impartial Sources"; Jon Cohen, "John Crewdson: Science Journalist as Investigator," *Science* (November 15, 1991): 946–49.

13. Nancy S. Wellman et al., "Do We Facilitate the Scientific Process and the Development of Dietary Guidance When Findings from Single Studies Are Publicized?" *American Journal of Clinical Nutrition* 70, no. 5 (1999): 802–5; David F. Ransohoff and Richard M. Ransohoff, "Sensationalism in the Media: When Scientists and Journalists May Be Complicit Collaborators," *Effective Clinical Practice* 4, no. 4 (2001): 185–88.

14. Boyce Rensberger, "Reporting Science Means Looking for Cautionary Signals," *Nieman Reports* (Fall 2002): 11–13.

15. Bob Beale, e-mail message to author, May 18, 2004.

16. Kovach and Rosenstiel, *Elements of Journalism,* 108.

17. Timothy Crouse, *The Boys on the Bus* (New York: Ballantine, 1972); John Burgess, "The Pottery Deficit Revealed: U.S. Closely Guards Trade Figures That Could Move Markets," *Washington Post,* April 20, 1991, B1; George Blake, "Are Embargoes Sacrosanct?" *ASNE Bulletin* (October 1984): 18–19; Tony Case, "*USA Today* Breaks Embargo," *Editor and Publisher* (August 15, 1992): 7; Price Colman, "New Olympic Event: Fighting over News Embargoes," *Broadcasting and Cable* (July 29, 1996): 8;

Debbie Creemers, "Are Embargoes Sacrosanct?" *ASNE Bulletin* (October 1984): 18–19; Phil Rabin and Carolyn Myles, "Is 'Embargoed' Like the Labels Reading 'Don't Open Till Xmas'?" *Washington Times,* April 20, 1994, B8; Jane A. Welch, "News Embargoes," *Presstime* (August 1985): 14–15.

18. Leonard Downie and Robert G. Kaiser, *The News about the News: American Journalism in Peril,* 1st ed. (New York: Alfred A. Knopf, 2002), 55.

19. Kovach and Rosenstiel, *Elements of Journalism,* 76.

20. Peter Gorner, "The Edgy State of Science," *Chicago Tribune,* February 13, 2000, sec. 2, pp. 9; National Association of Science Writers, *Hold for Release.*

21. Comments by Peter Wrobel to the National Association of Science Writers, Denver, February 13, 2003; Alexandra Witze, telephone interview by author, April 27, 2004.

22. Ibironke Lawal, "Scholarly Communication: The Use and Non-use of E-print Archives for the Dissemination of Scientific Information," *Issues in Science and Technology Librarianship,* no. 36 (2002).

23. Craig W. Trumbo et al., "Use of E-mail and the Web by Science Writers," *Science Communication* 22 (2001): 347–78; Shearlean Duke, "Wired Science: Use of World Wide Web and E-mail in Science Public Relations," *Public Relations Review* 28, no. 3 (2002): 311–24; Rebecca Dumlao and Shearlean Duke, "The Web and E-mail in Science Communication," *Science Communication* 24, no. 3 (2003): 283–308.

24. Jerome P. Kassirer and Marcia Angell, "Prepublication Release of Journal Articles," *New England Journal of Medicine* 337 (1997): 1762–63.

25. National Academy of Sciences, "Information for Authors," April 2003, http://www.pnas.org/misc/iforc.shtml (accessed August 15, 2003); Denise Graveline, "ACS Embargo Policy," *Science,* no. 282 (1998): 1643.

26. Jerome P. Kassirer, "Posting Presentations at Medical Meetings on the Internet," *New England Journal of Medicine* 340, no. 10 (1999): 803.

27. Tom Siegfried, telephone interview by author, April 2, 2004.

28. Witze, telephone interview by author.

29. Mike King, "Stories on Major Medical News Can't Be Put on Hold," *Atlanta Journal-Constitution,* July 13, 2002, 9A.

30. John C. Merrill, *Journalism Ethics: Philosophical Foundations for News Media* (New York: St. Martin's Press, 1997).

Index

VINCENT KIERNAN is a senior writer at the *Chronicle of Higher Education.*

The University of Illinois Press
is a founding member of the
Association of American University Presses.

―――――――――――――――――――――――

Composed in 10.5/13 Adobe Minion
with Meta display
by Jim Proefrock
at the University of Illinois Press
Manufactured by Thomson-Shore, Inc.

University of Illinois Press
1325 South Oak Street
Champaign, IL 61820-6903
www.press.uillinois.edu